Martian Linguistics

Images of Elsewhere

Vol. III

PETER LANG

Oxford - Berlin - Bruxelles - Chennai - Lausanne - New York

Martian Linguistics

Timothy Jenkins

PETER LANG
Oxford · Berlin · Bruxelles · Chennai · Lausanne · New York

Bibliographic information published by the Deutsche Nationalbibliothek.
The German National Library lists this publication in the German National Bibliography;
detailed bibliographic data is available on the Internet at http://dnb.d-nb.de.

A catalogue record for this book is available from the British Library.

Library of Congress Cataloging-in-Publication Data

Names: Jenkins, Timothy, 1952- author.
Title: Martian linguistics / Timothy Jenkins.
Description: Oxford; Bern; Berlin; Bruxelles; New York; Wien: Peter
   Lang, [2025] | Series: Images of elsewhere; vol. III (3) | Includes
   bibliographical references and index.
Identifiers: LCCN 2024035332 (print) | LCCN 2024035333 (ebook) | ISBN
   9781803741734 (paperback) | ISBN 9781803741741 (ebook) | ISBN
   9781803741758 (epub)
Subjects: LCSH: Life on other planets. | Extraterrestrial beings—Language.
   | Interstellar communication.
Classification: LCC QB54 .J46 2025 (print) | LCC QB54 (ebook) | DDC
   999—dc23/eng20240809
LC record available at https://lccn.loc.gov/2024035332
LC ebook record available at https://lccn.loc.gov/2024035333

Cover image: Line drawing by the author.
Cover design by Peter Lang Group AG

ISBN 978-1-80374-173-4 (print)
ISBN 978-1-80374-174-1 (ePDF)
ISBN 978-1-80374-175-8 (ePub)
DOI 10.3726/b20809

© 2025 Peter Lang Group AG, Lausanne
Published by Peter Lang Ltd, Oxford, United Kingdom
info@peterlang.com - www.peterlang.com

Timothy Jenkins has asserted his right under the Copyright, Designs and Patents Act, 1988,
to be identified as Author of this Work.

All rights reserved.
All parts of this publication are protected by copyright.
Any utilisation outside the strict limits of the copyright law, without the permission of the
publisher, is forbidden and liable to prosecution.
This applies in particular to reproductions, translations, microfilming, and storage and processing
in electronic retrieval systems.

This publication has been peer reviewed.

Contents

Series Preface — vii

Introduction — 1

CHAPTER 1
Yearning for contact — 3

CHAPTER 2
Communication and its failures — 45

CHAPTER 3
Exercises in Martian linguistics — 71

CHAPTER 4
Four novels of 'first contact' — 119

Bibliography — 145

Index — 151

Series Preface

Reports of flying saucers – also known as UFOs – constitute a puzzle, for they are numerous, well attested, and hard to believe. There are tempting shortcuts to a 'solution' – that the sightings are real, or mistaken, or fictitious (made up) – but none of these prove satisfactory. Instead, we are brought to consider the history of sightings and the history, also, of how it became possible to regard such incidents in the terms that have become customary. Flying saucers in this fashion become a feature of the wider society, and allow an angle of approach to our modern, technological civilization: a small-scale problem that allows insight into the larger setting.

The six essays stand as independent studies. Each deals with an aspect of the life of flying saucers or UFOs: their appearance after the Second World War within the constellation of military and technological interests, their debt to early science fiction and its sources, the development of the search for signs of extra-terrestrial intelligence, the first adoptions of the 'interplanetary hypothesis' in civilian circles, the further expansion of reports, first, of sightings and, then, of abductions in the wider society, and, finally, a review of the range of forms which have appeared. Taken together, they form a thorough enquiry into reports of sightings of flying saucers.

The series as a whole makes three contributions to resolving the puzzle posed by such reports.

First, it relates three bodies of materials from the United States in the mid-twentieth century whose interactions must be taken into consideration when speaking about flying saucers. These are the science fiction milieu, the interplay of military and technical interests, and reports of sightings by members of the public; in short, stories, military work, and ordinary lives. The first contribution is to study their interactions, overlaps, borrowings and synergies.

The second is to derive the categories that are necessary to explain the convergence of these materials. Repeating patterns appear in science fiction literature, the history of Air Force intelligence in the Cold War period, the early days of NASA, the search for extra-terrestrial intelligence, and a wide variety of incidents and claims made by members of the public. To make sense of their common nature and to see how their interactions work, we also need to investigate some intellectual history. There is a longstanding tradition of popular thought putting new scientific discoveries and technological innovation to work for human moral purposes. This tradition was taken up by military and technical interests in the middle third of the twentieth century, using three clusters of ideas: the intimate connection between military technology and the world picture offered by modern media, the concept of 'communication' (and, post-War, of 'information') that became central in the period, and an understanding of 'memory' as an exact record of the past. These ideas were shared with a wider public: in the context of international tensions, hopes of communication and fears of its breakdown were given expression in the appearance of new forms of life, forms given content by the earlier longstanding history. This is the second contribution the essay makes to the topic: an investigation of the common patterns of thought necessary for stories, military work and ordinary lives to interact.

And, last, a mechanism is proposed by which these interactions occur. This is an analysis of the ways in which these 'images', which contain both real and imaginary elements, make their appearance compelling. I find well documented instances – in particular, the sessions in which memories of abductions are recovered – where the social mechanism is uncovered that allows the oscillation between the two elements, a mechanism that can be glimpsed at work in other sites but which cannot be tracked in such detail in the documents and other sources we have concerning advances in research, security decisions, the records of incidents and so forth. This is the third contribution.

I first came to the puzzle of flying saucer reports when working on spirit messages and similar forms of social life (such as parapsychology and psychical research) and realized that the search for extra-terrestrial intelligence was the latest expression of a long-held desire for communication with disembodied minds compatible with our own. It has taken a good deal of time and work to give substance to this insight. As will be clear from my references, there is an abundance of work of the highest quality in this broad area, on which I draw to give shape to the argument. If I have contributed anything, it is by making a systematic enquiry and by putting together materials that are not always associated, and by continuing to ask questions rather than settling for accepted answers. In this fashion, I hope to have supported readers who find these topics interesting rather than those who wish to close them down, and I also hope to have contributed in some small degree to understanding the contemporary world.

Introduction

The interplanetary hypothesis – the idea that flying craft may come from other planets – arose within the American Air Force in the immediate post-War period.[1] But if tracing the beginning of the interplanetary hypothesis represents a first stage in the argument, we cannot stop there, for the image of flying saucers or UFOs continued its semi-independent life within the military-technical constellation. This essay begins with two case studies: on the one hand, the closing of the Air Force's Project Blue Book, dedicated to researching the UFO problem, and, on the other, the initiation of the search for extra-terrestrial intelligence (SETI) in NASA. The first serves to introduce this essay's principal topic: the potential for communication with other planets and its disappointments. The discussion of 'communication' and of the closely related concept of 'information' is central to the development of the argument. The second case study provides the main subject matter for that discussion. Together, these two studies point to a specific, historically located, understanding of the idea of communication and what it implies and excludes. The second chapter reviews the concept of communication; this recent perspective, in our reading of some relevant texts, has two features: it binds attempts to imagine communication with other planets ('Martian linguistics') within the limits of the ambition of mind-to-mind contact, on the one hand, and the impossibility of escaping our human world, on the other; and it summons up spirit figures to cross the divide, to act as relays joining other places and times. In two further chapters, we look at examples of how communication with other planets has been conceived within this framework over the twentieth century, considering both fictional and non-fictional examples of languages taken as means for communication between species from different planets. The non-fiction examples concern designs for interplanetary languages within the SETI project. The

[1] The argument is laid out in volume I of the series, *Flying Saucers: An Introduction.*

last chapter reviews four novels, from the late 1960s to the early 2000s, which deal specifically in imagined 'first contact' within the setting of the SETI programme.

CHAPTER 1

Yearning for contact

How do our conceptions of life in the Universe take form? How do they gain their characteristics? I want to look at a specific problem, the importance of the concept of 'communication' to space exploration, both fictional and real. I have some instances of how communication between human and interplanetary beings has been imagined, examples which offer insight into the presuppositions underlying our hopes of inter-species contact. Before coming to these instances, however, the context needs framing, for there is a striking history of events on the ground.

We start from two case studies. In the first place, I trace the story of how the United States Air Force rid itself of Operation Blue Book, its intelligence unit concerned with flying saucer sightings, and sought to control both the reporting of sightings and the reception of those reports, aiming simultaneously to damp down public interest in flying saucers and to extricate itself from involvement in the topic. The first framing was then in terms of intelligence concerns: were flying saucers evidence of another power gathering sensitive information? What were their technical capacities, and could we match their achievements? And secondarily, what were their intentions and what should our response be? Cooperation or competition? As no clarifications emerged, however, these questions quickly gave way to a quite different set of concerns, about the reception of the reports and their effects on the population in the context of the Cold War: fears that flying saucer scares might distract attention at vital moments of a potential Soviet attack and fears, too, of their serving as a channel for disloyalty and subversion, a betrayal of American interests. Having inadvertently hatched the idea of flying saucers, the Air Force was then forced to meet the reactions they provoked, and the path it went down was first to marginalize and then eliminate consideration of reports of sightings. We shall follow this process in the first case study.

In the second place, in the same period (1950–1970) the Space Programme developed and included in its interests the Search for Extra-Terrestrial Intelligence (SETI). This provides the second case study, which is the main focus of this essay. In short, by 1970, military interest in flying saucers and their possible significance formally had been ended, yet, in the Space Programme, new efforts and budgets were being expended on finding signs of extra-terrestrial civilizations, without any sense of tension, contradiction or paradox. The starting point of the argument is the seeming conflict between the two cases – the attempted closing down of questions about interplanetary visitors, on the one hand, and beginning the search for interplanetary communication, on the other. The resolution, if that is the right word, is that both are instances of disturbances in perception produced by technological innovation, disturbances making their appearance in the collective representations by which the organizations concerned simultaneously understood, promoted, and explained these innovations. Both the military and the space programme, as we shall see, in fact follow the same trajectory, first producing positive and, later, more negative evaluations of these modes of representation.

The central concept of this process of representation turns out to be that of 'communication', conceived both in terms of hopes and failures, anticipation of new contacts and disappointment of those expectations. And, within this concept, the contemporary idea of 'information' plays a particular role in its realization. These two notions – communication and information – allow us to understand how the disturbances in underlying categories brought about by technological innovation are grasped and made sense of: how they are realized in perceptions which promote certain elements of the picture and miss or distort others.

The common ground to the two case studies – the 'solution' to the puzzle – lies then in the concept of communication and its several aspects, which therefore deserve examination. I pursue this discussion in the second chapter, for it is indeed a keystone to the whole range of arguments concerning flying saucers. Once we have identified the importance of these organizing categories, we can turn to consider the characteristics that mark the study of 'Martian linguistics', which is a way of summing up the project of interplanetary communication.

In this fashion, we explore the later history of the interplanetary hypothesis – its elimination from one institutional zone and appearance in another – and review the terms in which contact between planetary civilizations is imagined, both in fiction and in reality. We are concerned with how the image which emerged in the early years after the end of the War in 1945 was put to work subsequently: how it is 'used to think with' while simultaneously exhibiting a life of its own, acting autonomously in more than one complex institutional environment. The focus on the idea of communication, then, allows us to explain features of SETI in the context of the far larger project of space exploration and to trace a series of imaginative uses of the original image.

We have to take a distinct approach in this chapter in setting out the two case studies. Rather than focussing on a single text from the period to detect traces of shifts in perception and understanding, we have recourse to syntheses in order to sketch the development of a series of attitudes in two large institutions, the Air Force and NASA, over a decade and more. I seek to identify switch points in these histories, when new projects were first conceived and then hatched and note, too, when such projects were terminated. But the initial descriptions are broad-brush and rely in large part on the judgements made in existing histories. They open the way to some intellectual history in the second chapter that introduces the organizing theme of the essay, the centrality of the concept of 'communication' to a series of modern projects including the involvement of the Air Force with sightings of flying saucers and the Space Programme's engagement with the possibility of extra-terrestrial intelligence. With this theme in mind, we turn to the business of close reading of specific texts in their setting in the last two chapters, looking at the interplay between fictional and non-fiction descriptions of communication with other forms of life. In short, we outline a puzzle, identify a solution, and explore its out-workings in a series of illustrations and exemplifications. I have split the argument concerning examples, making the final chapter a review of four novels dealing in interplanetary contact.

I. The Air Force story

The first case study: in 1969, the Air Force closed its unit concerned with collecting and investigating reports of UFO sightings, Project Blue Book, following an Air Force funded university research project on UFOs, the Colorado project, which had issued the Condon report (named after the project director) the previous year.[1] The Air Force conclusion, stated in a memo of October 1969 on the basis of the report, was that 'the defense function [of protecting national security] could be performed within the framework established for intelligence and surveillance operations without the continuance of a special unit such as Project Blue Book' (cited Swords and Powell 2012: 336).

The Air Force position had been set by the Robertson panel findings in 1953, defined by Swords as a 'restrictive information management strategy' (Swords and Powell 2012: 204), with all reports and enquiries being channelled through the Public Information Office in the Pentagon. A typical sequence then consisted in an incident – a sighting and a report – followed by 'a non-committal response from local military authority, a subsequent debunking response from the Pentagon, and no other authority commenting'. In short, the increasingly effective imposition of an official line – perhaps more widely characteristic of the period – continued with the assumption that, in principle, every sighting could be explained by natural or known causes. As a result of this policy, Swords suggests, the initiative passed elsewhere, to the continuing rhythm of sightings, both military and civilian, and to the various civilian organizations which had emerged that made these reports their business.

This interpretative frame, focussing on information management rather than investigation of sightings on the part of the Air Force, noting the rise of independent reports of sightings, and investigations of reports passing to civilian hands, was set by Jacobs and is adopted by Swords, who supplements Jacob's account with more recently released documents. This

---

[1] I shall follow in outline Swords' account (Swords and Powell 2012), which in turn is shaped by Jacobs (1975).

approach reproduces in large part the presuppositions held by the civilian organizations, which had in turn trapped the Air Force and effectively determined its options.[2] The history is therefore told within a particular optic and reinforces the argument that state organizations are by no means always masters of their actions nor, even, their categories. Nor can we expect fully to escape that optic; at the least, we are bound to observe the categories of the various actors.

There are two crucial years in the history that interests us between 1953 and 1970: 1957 and 1968.

Sputnik – 1957

The launch by the Soviet Union of the first Sputnik on 4 October 1957 was, from the perspective of the history of flying saucers, an accidental event with extensive effects. It was the first man-made satellite to orbit the Earth and was followed by the much larger Sputnik II on 3 November. These technical achievements emerged in the context of an accelerating competition to develop supersonic aircraft and missiles capable of delivering nuclear warheads and represented a perceived crisis in American ambitions to be ahead in the arms race. In response, military budgets devoted to missile and space research were expanded and the financial means voted by Congress; the technological and intelligence communities created the Advanced Research Projects Agency (ARPA, later DARPA), formed in 1958 to develop advanced weapons systems and 'prevent technological surprise' (Swords and Powell 2012: 238); and, at the same time, the National Aeronautics and Space Administration (NASA) was created out of the existing National Advisory Committee for Aeronautics (NACA), formally coming into existence in October 1958.

With regard to Project Blue Book and Air Force provision more widely, the policy set by the Robertson panel continued, assisted by a relative drought in reports received. This was part of a long quiet period in terms of sightings which ran from 1952 to 1964, with the exception of a single

---

2   This account is outlined in the first essay.

spike in November and early December 1957, coinciding with the launch of Sputnik II. In this brief moment, sightings of a new kind were reported – 'Close Encounters' – signalling a shift both in the proximity of the craft and in the kind of witnesses, who were drawn from the civilian population rather than the military. Symptoms of Close Encounters included the failure of machines and equipment – car engines, lights, radios, even radar apparatuses – coinciding with the presence of the objects, and also some evidence of mild 'burns' to individuals, which could indicate a radioactive source. There was speculation on whether vehicle transmission failure pointed to 'some type of anti-gravitational control' (Swords and Powell 2012: 263), and there was interest from government laboratories in radar interference because of the possibility of developing a non-lethal battlefield tool (Swords and Powell 2012: 252). The intense military and political interest engendered by Sputnik found an echo in public sightings; with the launch of satellites visible from earth, everybody was looking at the skies.[3]

Civilian organizations concerned with UFOs had developed after Robertson and one, NICAP, the National Investigations Committee on Aerial Phenomena, was a particular beneficiary of the new situation, for it was both timely and well-located: it was a group formed in Washington D.C. in October 1956. Donald Keyhoe, a writer who had offered published accounts of Air Force strategy with respect to flying saucers from 1949 and who had played a major role in establishing the category of flying saucers in the public mind (see Jacobs 1975: 56–57), rose rapidly to prominence; he first recruited a series of senior retired military men for the board of NICAP and then, with their support, replaced Townsend Brown, the first director, in January 1957. Under Keyhoe's leadership, the organization focussed on forcing a policy change in the Air Force's handling of reports by lobbying politicians as the source of military policy and funding. Keyhoe's ambition was to make the Air Force disclose its 'hidden' records and, in this way, to expose the high level 'conspiracy of silence' which organized their actions concerning flying saucers (see Jacobs 1975: 145–147). In essence,

---

3   The Smithsonian Astrophysical Observatory organized Operation Moonwatch, under the direction of Fred Whipple and assisted by J. Allen Hynek, using amateur astronomers to plot US and Soviet satellite orbits (Swords and Powell 2012: 249f.).

he construed the shift in Air Force interest from investigation of reports to controlling public reactions to them, embodied in the Robertson panel report, as indicating the concealing of hard evidence and, possibly, of a history of contacts. His focus became the contest with the military to make them reveal what they supposedly knew and the story of their supposed evasions and tactics; he lost sight of the original interest in flying saucers, to which he had made an early contribution. In both regards, he was influential on the later ufological movement and set the broad premises of both Jacobs and Swords' approach.

Swords claims that Operation Blue Book had already responded to Keyhoe's attacks on military secrecy before the founding of NICAP, suggesting that the Air Force published the statistical report on filed cases commissioned by Edward Ruppelt from the Battell Memorial Institute in part to counter Keyhoe's release of new cases (in Keyhoe 1953). This counter was launched as *Blue Book Special Report 14* in 1955 (Project Blue Book 1955), reviewing the records and being spoken of as the 'death knell of saucers' (Swords and Powell 2012: 221), emphasizing the small proportion of 'unidentified' cases.

The report was reissued in 1957, in the atmosphere of increasing Congressional interest, and distributed to politicians. It was a shortened version, with a new preface and additional materials, bringing the evidence up to date. The new preface emphasized the 'improvement in reporting, investigating, and analytical techniques' (cited Swords and Powell 2012: 240), and hence the diminishing importance of unexplained cases. A chart was presented which suggested that press interest over time (articles and press releases) promoted the frequency of civilian reports. The text also noted the existence of 'detractors' of the Air Force. In Swords' view, the revised text was 'entirely fixated on problems resident in the public rather than those resident in the phenomenon. It was as if, for Blue Book, the phenomenon had disappeared' (Swords and Powell 2012: 239).[4] In this regard, it matched Keyhoe's contemporary focus.

---

4   This perspective is found in Jacobs (1975: 139ff.) and also confirmed in the earlier account by Hynek (see Hynek 1972: 189–215, particularly 202ff.).

In the addendum to the report, material was repeated from the testimony of the Director of the National Advisory Committee for Aeronautics before the House Appropriations Committee in February 1957, who 'flatly denied the existence of space vehicles' and, when asked why scientists did not make such declarations more commonly, replied 'we would have no time to do anything else. We cannot compete with the science fiction people' (cited Swords and Powell 2012: 241). When challenged by NICAP to produce the evidence for such a statement, he had to reply that he had none; it was simply personal opinion.

There was therefore tension between NICAP and the Air Force prior to the launch of the two Sputniks and the brief wave of Close Encounters which coincided. The combination favoured NICAP's tactic of seeking a Congressional hearing on Air Force handling of reports. According to Jacobs, the Air Force made a major objective between 1957 and 1964 of preventing or limiting Congressional hearings that touched on their handling of the UFO problem; this was with the declared aim of avoiding provoking a public scare and of having to declassify files, and with the further end of not revealing past alterations in policy (see Jacobs 1975: 158). They also invoked issues around betraying the privacy of witnesses and revealing classified knowledge concerning the capacities of electronic equipment (Jacobs 1975: 136–137). The Air Force's fundamental problem was that, while their critics' aim was to lay bare their supposed secret knowledge and manoeuvres, their practice since 1953 had been to focus on containment, not research. On the one hand, they had no secrets to reveal of the kind being sought, while, on the other, making this clear would also make clear the distrust of the ordinary man and woman that lay behind the policy of fearing public reactions and the possible manipulation of their responses.

NICAP made a series of attempts to obtain Congressional hearings, frustrated by Air Force countermeasures; Swords offers a list of these (Swords and Powell 2012: 275ff.). Their nearest approach to success came in the period we are discussing when a sub-committee of the House Select Committee on Astronautics (*sic*) and Space Exploration – the sub-committee on Atmospheric Phenomena – held meetings on the subject of UFOs in August 1958. The members received materials submitted by the Air Force and discussed reports and investigations, were shown photographs and films, and reviewed the role of 'citizen clubs, books, and organizations' and their effects on Air Force responsibilities (Swords and Powell 2012: 278). The intention was expressed to interview witnesses from the public as well, which would have offered NICAP the opening they sought, but this possibility was not followed up for reasons that are not clear.

Swords offers evidence of attitudes within the Air Force to this repeated pressure, citing a memorandum sent from ATIC (Air Technical Intelligence Center, which included Blue Book) to Pentagon Air Force Intelligence in December 1960. It notes that there are 'more than 50 private unidentified flying object (UFO) organizations', boasting more than half a million members; it describes their principal claims as, first, that UFOs are 'interplanetary visitors' and, second, that 'the Air Force is withholding information it has concerning them'; and speculates that members' motives for belonging were either financial, religious, emotional, from ignorance, or for Cold War purposes (Swords and Powell 2012: 292).

In short, by the early 1960s, the relevant elements in the Air Force were locked in a sterile struggle with NICAP who effectively dictated the terms of the argument, despite the fact that the latter's power for the most part derived from the Air Force strategy of focussing on managing public opinion on the topic of flying saucers. The press held the ring for this contest, with politicians alternating between the public need to respond to events while in private being constrained to act responsibly in agreement with the information provided by experts. As Swords notes, in this period, scientists for the most part derided the possibility of UFOs while at the same time pursuing the idea of what would become known as SETI (the Search for Extra-Terrestrial Intelligence) (Swords and Powell 2012: 297).[5]

Under these conditions, by 1961 the objective of those involved in Operation Blue Book had become to 'get the project eliminated or passed over to some other element of the Air Force or another institution (e.g.

---

5   Though there were plenty of exceptions. Swords cites a passage from a talk by Dr Hermann Oberth, 'Werner von Braun's mentor in rocket technology', who visited the States in 1958; he offers one of the best technical accounts of how the various sightings of flying disks might be integrated into a single description (see Swords and Powell 2012: 274–275).

NASA) entirely' (Swords and Powell 2012: 293). This objective was achieved in 1969.

The Colorado project – 1968

The Air Force position had become locked in frame; having switched from an early position of analysing the residue of cases that resisted explanation to the security-focussed desire to shape public responses to reports, and then confronted with organized lobbying projecting a narrative of secret knowledge and a strategy of obfuscation onto them, Operation Blue Book was caught in an impasse to which the only lasting solution was to drop any responsibility for receiving reports.

There were two motives for supporting this aim within the organization. The first was that the Air Force was not assisted in its primary task of ensuring national security by this distraction which was always threatening to return. This position was represented by Blue Book's last officer in charge, Major Hector Quintanella who, in Swords' account, saw his task as collecting data without any analysis and managing public reactions by negative comments (Swords and Powell 2012: 306). The other – a minority view represented by J. Allen Hynek, who had acted as scientific advisor to the project from its early days – believed the proper investigation of reports would only be pursued once it ceased to be solely Air Force business.

Once again, the motive power that allowed this reform to occur was provided by an increase in sightings and the wider response this evoked, in this case, a flap which began in 1964 and which came to public prominence with the Socorro incident in March 1965. In August 1965, Hynek proposed to the Pentagon the idea of involving the National Academy of Sciences, on the grounds that a panel of academics, both physical and social scientists, could assist the Air Force to resolve its combination of scientific and social problems. By the end of 1965, the Pentagon had decided to present the UFO problem to an ad hoc group of the Air Force Scientific Advisory Board; this group held a one day meeting to review Blue Book's work in February 1966 and 'recommended strengthening the investigation by contracting with an important university … to do in-depth research on one hundred or so sightings per year' (Swords and Powell 2012: 307). This proposal was then to hand when national politicians felt called to react to public disquiet over the official explanations offered in another case, the Dexter and Hillsdale lights in Michigan, in late March; the Secretary to the Air Force, Quintanella and Hynek were called to testify before the House Armed Services committee on 5 April, and the proposal for further research became a means for laying the dust that had been raised.

Responsibility for recruiting a university and setting up and evaluating a project fell to the Air Force Directorate of Science and Technology and, within that organization, the Office of Scientific Research. A chain of contacts, professional, personal, and financial (previous research grants), led to the University of Colorado becoming involved and the head scientist at Colorado, Edward Condon, becoming the project leader. Funding was agreed, a project set up, and scientists and administrators recruited.[6]

This project has been discussed a great deal by ufologists; it was controversial at the time and has been analysed repeatedly since.[7] Two factors appear important in the controversies. The first is that natural scientists have no agreed procedures for investigating phenomena such as flying saucers. There is a good deal of evidence to this end from minutes on project design and from an internal review conducted by the university. This difficulty lay behind the university's initial reluctance to become involved and resulted in a memorandum (the 'Low' memorandum) which became the subject of controversy and was originally part of selling the project – a

---

6   This chain is traced by Swords (Swords and Powell 2012: 307–312); he relies a good deal on the contemporary account found in Saunders and Harkins (1969), who were participants in the project.
7   See the summaries in Hynek (1972: 217–241), Jacobs (1975: 225–263), Peebles (1994: 169–195), Clark (1998: 592–607), Hoyt (2000), and Swords and Powell (2012: 306–335). Hoyt offers an analysis of Condon's past history and motivation, pointing out that he defended the autonomous values of scientific research both during the War and after, against persecution in the McCarthy era, and making the case that he extended the same motives to the Colorado project; he was opposed to the methods of ufologists rather than to the idea of UFOs as such (see Hoyt 2000: 68). On the mutual mistrust of natural scientists and ufologists as a more general motif, see Eghigian (2015).

political necessity to the Air Force – to the administration and Board of Regents at Colorado.

Second, although the project was conceived in the form of a scientific investigation, its purpose remained that of ridding the Air Force of the responsibility of investigating flying saucer reports. This was borne witness to by correspondence, early in the development phase, between the representative of the Air Force Directorate of Science and Technology, the sponsor of the project, and, at the university, Condon and Low, the project's administrator. Condon achieved this end by writing the report Summary and Recommendations, which were not 'congruent' (Swords' word) with the findings of the other sections, incorporating the view 'that no funding or ongoing research be facilitated' (Swords and Powell 2012: 330). By the time of publication, most of the scientists working on the project had broken ranks, with recriminations, resignations, and firings, and the report provoked a good deal of public controversy.[8] But the scientific establishment – *Nature*, the National Academy of Sciences, Whipple at Harvard, and others (Swords and Powell 2012: 331) – rallied, welcoming the report, and the Air Force got what they wanted, grounds to close Project Blue Book.

Because responsibility for Blue Book had been transferred to Air Force Research and Development Command, the memorandum recommending the closure of the Project was signed by the department's Deputy Director of Development who, by a nice conjunction, had managed the lunar module construction for NASA (Swords and Powell 2012: 336). As Swords makes clear, however, closing down the operation only cut out a small part of the already existing system for reporting non-standard sightings: 'sensitive intrusions' continued to be recorded; 'only a publicly-known office, a source of continual irritation, went away' (Swords and Powell 2012: 337). There was now no acknowledged site to maintain 'a large active file of non-military cases' (he speculates, following Hynek, as to whether militarily sensitive cases were ever passed to Blue Book), but whether systematic attention

---

8 Swords points to John Fuller's article in *Look*, May 1968, 'Flying Saucer Fiasco' and to an article by Boffey in *Science*, 26 July 1968 – see Swords and Powell (2012: 328–329).

was paid to such matters elsewhere in the military or security apparatus remains a matter of conjecture.

There is some evidence of continuing interest by the Air Force in UFOs; a clutch of documents relating to UFO sightings at Air Force bases in 1975 were released under a Freedom of Information request, which contained such familiar elements as lights, manoeuvrability, merging and separating units, multiple witnesses, radar and visual sightings, repeated visits, tampering with electrical systems, and interest in nuclear weapons storage areas (see Swords and Powell 2012: 338; Swords cites Fawcett and Greenwood 1984; Hall 2000).[9]

Swords sees the Colorado project and the end of Project Blue Book as marking some kind of watershed. Not only did the Air Force rid itself of any publicly acknowledged interest in UFOs, but the civilian organizations also went through some kind of metamorphosis. NICAP ceased to exist along with its object of desire, Blue Book, and APRO (the Aerial Phenomena Research Organization) faded at the same time. Ufological interests were recast with the emergence of the Center for UFO Studies (CUFOS), organized by Hynek, and the Mutual UFO Network (MUFON) began its growth to become the most prominent national UFO organization. At the same time, Swords also notes that interest in flying saucers became more acceptable among scientists,[10] and Hynek sought to create an 'invisible college' of qualified participants to join in his investigations (see Hynek 1972, Part III).

One further remark: the term 'extra-terrestrial' gained a new prominence in the course of the Colorado project and may be a marker of the shift Swords describes. Clearly, the word has a long history of non-technical

9   Further materials appear from time to time; there has been a release of CIA documents relating to 'unidentified aerial phenomena' (UAPs), including some film, during the writing of this essay – see *The Guardian* 13 January 2021. The article also mentions current Congressional interest in the topic, with national defence interests predominating.
10 He points to a symposium which led to the publication of *UFOs: a scientific debate*, edited by Page and Sagan, in 1972, and to 'UFOs: an appraisal of the problem' in *Astronautics and Aeronautics* in 1970. We can also note Kuettner's article in the same journal in 1973, entitled 'A new start to the whole UFO problem?'.

use: the term 'terrestrial' appears to be late sixteenth century in origin, and 'extra-terrestrial' appeared in the mid-nineteenth century. But it emerges as standard usage in the documents and circles we are considering, as exemplified in a position paper by the project administrator, Robert Low, in April 1967. He asked three questions: 'are there really sightings that cannot be explained?', if so, 'are any of these solid objects?' and, third, 'are any of these objects extraterrestrial spaceships?' (He also offers the alternative, that these may be 'terrestrial phenomena of an as yet unknown source and description') (see Swords and Powell 2012: 317f.). According to Jerome Clark, its specific use in the phrase 'the extra-terrestrial hypothesis' (later shortened to ETH), appeared in the project's final report: 'Condon invented the phrase, defining it as the "idea that some UFOs *may* be spacecraft sent to Earth from another civilization, or a planet associated with a more distant star"' (Clark 1998: 213). The term marks a shift in registration, with new players, new stakes and, to an extent, a different game.[11]

In the wider perspective we are exploring, the clash between the Air Force and NICAP represents a series of repeated failures of communication. From the ufologists' side, the Air Force was held to be ignoring the significance of incoming interplanetary craft or, more probably, to know a good deal about their agenda and, even, to be in touch with them but successfully concealing this exchange of information. From the Air Force side, their task may be defined as a series of attempts to control the ill effects of miscommunication: reports of flying saucers were held potentially to cause panic among the population, to aid the enemy by creating temporary blockages in the communications system (the telephone network) and, longer term, to undermine trust in the Air Force and so to be possible evidence of propaganda work by anti-American agents. Their means for controlling these contagious messages took the form of denial, counterpropaganda, and education of both adults and children, and was marked by a sense of being put on the back foot by the publication and dissemination of ideas by publishers, journalists, broadcasters, and the civilian organizations that

---

11   As a point of detail, 'ETH' is used by the Air Force contact in the private correspondence with Low and Condon (see Hoyt 2000: 46) and so appears to have been current shorthand, at least in the circle that defined the concerns of the final report.

promoted the idea of UFOs. The story is less that of communication than the failure or clash of projects of communication at many points.

We may observe that the notion of communication at stake in these anxieties and policies is, at best, approximate. The events are not defined by clear instances of information going astray and evaluation of the damage done (as may be estimated in cases of espionage), but rather by vague fears of possible effects of an ill-defined nature. Nor is there any clear sense of the mechanisms by which these effects are brought about, nor how they may be countered, only poorly defined threats and equally formless remedies. We are speaking of a moral vision given expression in institutional gestures, organized through a particular understanding of communication and its effects.

II. The National Aeronautics and Space Agency (NASA)

The second case study: My purpose in tracing one of the aspects of the creation of NASA is to elicit the reasons for a focus on, and confidence in, the possibility of communication with other civilizations. This has never been a major feature of NASA's programmes but, nevertheless, has been a repeated minor theme. In seeking to elucidate this element, I shall rely in the main on William McDougall's 'political history of the Space Age', ... *The Heavens and the Earth* (1997, first published in 1985).

McDougall identifies a series of decision points in the history of the development of the American Space Programme. He places that history in the context of post-War military priorities and, therefore, in relation to the evolution of the Cold War with the Soviet Union. The creation of military technology in response to the threat of armed conflict effectively came to control the national agenda, and technological considerations overrode other political, economic, or cultural values. In this regard, McDougall confirms Kittler's thesis (see Kittler 1987), although he adds considerable nuance in matter of detail, in particular describing American politicians' reluctance to become fully committed to state-sponsored military research and development in the years immediately following the War. The setting up of the space programme represents the triumph of what he calls 'technocracy', a step in the social embodiment of some longstanding trends.

In the term 'technocracy', McDougall includes such ideas as 'the management of society by technical experts', as well as 'the institutionalization of technical change for state purposes' (McDougall 1997: 5). Later,[12] he speaks of 'technology ... drafted into the service of political agendas' and expands the notion to cover 'the dream of limitless progress through government-sponsored research & development' (McDougall 1997: xv). He traces a genealogy of the novelty from the First World War, which developed a 'model of command economics in technology, as well as in investment and distribution generally' (McDougall 1997: 5), and he sees it coming to fruition in the 'command technology' of the Second War, which produced four outstanding technological advances: radar, atomic weapons, ballistic missiles, and the computer (McDougall 1997: 6). This model was bequeathed to the post-War period and set terms not only for military developments but also for imagining a civilian agenda which would confront such social problems as poverty, health, housing, education, transportation, and communication, in this way eliminating want and so (potentially) removing the material causes of war.

He suggests that Russia took the lead in developing the idea of technocracy, even in its pre-Communist history, combining an ideology of the importance of science and technology to the task of modernizing a 'backward' society with an autocratic form of government, so that technological and military progress formed a great part of the legitimation of those ruling the state. This 'settlement' was not altered by the experience of the Second War so that, confronted with an American rival with marked technological superiority, in particular, the possession of atomic weapons, the Soviet response was to put its state-organized research and development into a competition to equal and surpass America's military achievements. McDougall's initial thesis is that Russia, as the world's first technocratic state, determined the move to 'Cold War'; that, in this regard, technology determined the form politics took (see McDougall 1997: 47).

---

12   In the preface to the second edition, McDougall (1997).

Seen in retrospect from 1997, he can suggest that 'the Space Age seems almost coterminous with the Cold War itself' and offers the following summary:

> That age was born in the initial competition between the Americans and Soviets to get their hands on Nazi V-2s and their designers. It accelerated in the 1950s as both sides raced for an intercontinental ballistic missile. It took off with *Sputnik-I*, climaxed with the Moon race, declined with détente, and died when the Soviet Union died. (McDougall 1997: xvi)

We shall see the interest of this claim to our narrower concerns in due course. Part of the story, however, is how the United States, despite its technological superiority, only fully adopted the command model of defence funding with the opening of the space race in the mid-1960s. The US had a long tradition of government staying away from funding science as the means of generating economic wealth. Private enterprise and individual initiative produced new tools and methods, and the state might respond to these independent inventions by taking up some for its purposes. In this regard, McDougall sees the American experiments in technocracy conducted in the Wars as limited and, in principle, temporary expedients; the development of atomic weapons and radar were under military control in wartime and, after 1945, weapons research was continued in-house in Army and Navy arsenals and the Atomic Energy Commission. The overall aims in peacetime were to control budgets, releasing money for social projects, to free industry and enterprise from state restraint, and, more generally, to protect civil society from the secrecy and control necessary for centralized planning. The contrast between America and Russia was thought in the period in terms of an opposition between 'Open' and 'Closed' societies.

There was one long-term exception to the rule concerning the lack of interest by government in research and development, where the state took primary responsibility for generating new technology, which was the National Advisory Committee for Aeronautics (NACA), formed in 1915 (McDougall 1997: 75), which later offered an element which could be developed into a state-backed civilian agency dedicated to space research (cf. Bilstein 1989). And the increasing importance of aerial warfare in the

1940s, recognized in the separation of the Air Force from the Army in 1947, indicated the direction that defence needs would take; the Air Force was exceptional among the services in its cooperation with independent contractors both in the development of weapons and the appraisal of the international situation, contracting with corporations and think tanks.

The main objective of government after the War was then to balance the need for adequate defence against the interests of the home economy (McDougall 1997: 73). Yet, in the context of a rival power seeking to match and surpass American expertise, in particular in nuclear weapons and delivery systems, there could be no return to traditional peacetime conditions, with the military being held in reserve; armaments and weapons development needed to be maintained in readiness. The initial compromise was a policy of containment and deterrence, developing long range bombers and making and storing atomic weapons.

The Korean War

This balance was shifted further in the direction of a technocratic 'solution' by the outbreak of the Korean War. Two elements emerged in this context that shaped the future space programme: the need to match Russian advances in rocketry, which led to the race to create intercontinental missiles, and the equal need to obtain direct aerial intelligence on Russian innovation and preparedness. Each of the armed services had its own rocket projects, while the need for aerial intelligence led to research in high level atmospheric flight (stratospheric balloons and supersonic aircraft) and satellites (see McDougall 1997: 97). The first stimuli leading to space flight were exclusively military.

The first conceptual review of the potential of space technology was undertaken by the RAND Corporation in cooperation with the Air Force, which produced a document in October 1950 which McDougall sees as 'the birth certificate of American space policy' (McDougall 1997: 108). This report already goes beyond purely military questions to consider wider questions of the reception of technological achievements. It discussed the value of a US satellite programme in both peace and war. Satellites (which would not exist practically for another seven years) will be developed for information gathering, allowing both 'strategic and meteorological reconnaissance'. They will also, however, raise questions of perception – of prestige and propaganda, we might say – so that their 'politico-psychological effects' must be considered. Their launch cannot be kept secret. In the context of the accelerating competition between the two countries and, in particular, the drive to produce Inter-Continental Ballistic Missiles (ICBM), the best policy would be 'to downplay the military potential of satellites and … [to stress] the peaceful aspects of this "remarkable technological advance"' (McDougall 1997: 109). The Americans should emphasize that 'the satellite was not in any sense a weapon' and, rather than relying on secrecy, should aim to control advance publicity. The strongest case for legitimacy was to show the satellite collected scientific data that could be shared and, in this fashion, to establish a precedent allowing flight over another nation's territory; military purposes could follow (see McDougall 1997: 110).

In sum, the document proposed disguising the military imperatives behind the technology by emphasizing the peaceful, civilian possibilities it also promoted. This was in response both to matters of presentation at home and internationally, with allies and rivals. It explored the issue of 'freedom of space' and, in brief compass, raised the contrast between an open and a secretive society; America had need to resort to direct surveillance to penetrate Soviet secrets, while the Russians could discover what they needed to know from more open sources.[13]

This report led, first, to a technical study involving a number of corporations under the sponsorship of the Department of Defense's Research and Development Board and then, in 1955, to the USAF commissioning a project to produce a 'strategic satellite system', integrating advanced technology from 'a dozen fields of American industry … It was a paragon of peacetime command technology … and the first American space program' (McDougall 1997: 111).

This revised holding position – a developing ICBM programme, a developing satellite programme – which held from 1950 to the middle

---

13 We might note the parallels with the contemporary American programme for the Peaceful Atom (McDougall 1997: 127).

years of the decade (see McDougall 1997: 123f.) was thrown in turn into disarray by the successful launch of the Soviet Sputnik I in late 1957. This led to a further and decisive step in the creation of NASA, the National Aeronautics and Space Administration, in 1958.

The Eisenhower years

In McDougall's words, 'American Space policy dates from the last Eisenhower years' (McDougall 1997: 140). He recognizes Eisenhower, who was President from 1952 to 1960, above all as the figure caught at the hinge between the desire to maintain a recognizable American peacetime civil society and the demands of a growing technocracy. Eisenhower had aimed at sustaining a peacetime separation between political interests, university and industrial research and development, and military projects, in order to protect citizens from a powerful technocratic state, dominated by private interests and justified by secret concerns.[14] Yet this position was bound to lose ground, and the launch of the Russian Sputnik was the precipitating point for a series of changes, accelerating research, demanding greater budgets for military projects, leading to a reordering of institutional relationships and, at the same time, the creation of a new relation to the public, so that the new conditions, with their expense, sense of threat, and shift in democratic relations, might be justified in the light of higher aims than simply responding to the achievements of a rival power.

Once again, the weaker side set the terms of the response. Just as the Soviet focus on state-driven military and technical progress pushed the US towards a centralized, technocratic response, so in the late 1950s the Russian emphasis on the cultural significance of its technological achievements determined the ground and some of the terms of American initiatives. The Soviet claim was that the technical achievements of the first satellite

14 In this regard, he shared the fears of ordinary citizens to which pulp writers had given highly coloured expression a decade before – see the second essay, *Religion and Science Fiction*.

launch posed a cultural challenge, serving as a symbol of the hopes offered by Socialism to the newly independent countries freed from colonialism. From the American side, the space programme could not simply be a means of focussing a technical response to the military threat implied by having the rocketry to launch an orbiting space vehicle. From an inside point of view, priority in the first successful satellite launch was of little significance; American investment and progress in space research was generally at a higher level than Russian industry and research could sustain. But in terms of public perception, Sputnik was a significant moment in a competition for prestige between the American way of life, based in an image of free enterprise and capitalism, and the socialist way, with state planning underwriting technological progress. Sputnik served as a way of summarizing the promise contained in each, not only for the newly independent nations but also for the domestic population. National debates about the role of education, the funding and quality of scientific research, and the place of racial minorities (echoing the status of the newly free nations) became recontextualized in this period.

The launch of Sputnik then caught the evolving rocket and satellite programmes in a different frame of reference. McDougall details the political compromises that followed and resulted in the National Aeronautics and Space Act which, late in 1958, created a civilian agency, NASA. Dean (drawing on McDougall) offers a summary of the significance of this civilian status:

> For ... many in politics and mainstream new media, America's space program had to reflect American ideals. If Soviet efforts in space were military secrets, then America's would be open and public, a civilian operation in the interests of peace. The US program had to appear ... [to be] pursued for the sake of "freedom in space". The decision to house American space efforts in a civilian agency, then, was directly linked to America's political goals against the Soviets: only a civilian agency could convey this open, peaceful image. (Dean 1998: 71–72)

With a civilian agency came a preoccupation with communication, as Dean insists; a civilian agency signals the weight that had to be given to political considerations – elements of prestige and presentation – equal to the concern with defence. The agency's tasks were not only linked with convincing peoples of other countries of American technological prestige, but also with assuaging domestic doubts that had arisen concerning the state of American science research and education (McDougall 1997: 157) and creating a political project of a 'New Age beyond deterrence', which was given form under the Kennedy administration with the programme to land a man on the moon.

The new agency was built around NACA (the National Advisory Committee for Aeronautics), which had played a part in the post-War development of jet propulsion and supersonic flight. In McDougall's judgement, 'NACA was an adjunct, not a rival of the Pentagon and industry' (McDougall 1997: 165), and this made it a plausible candidate, given the twin imperatives of meeting defence concerns and a civilian status, signalling the space programme's separation from direct military control.

In effect, the National Aeronautics and Space Act of 1958 'chartered two parallel space programs, one open, scientific, and devoted to research, the other closed and devoted to military applications' (McDougall 1997: 172). Its subject was the newly formed agency, NASA, and it left Department of Defense responsibilities in the shadows, with the newly created ARPA (Advanced Research Projects Agency) coordinating military programmes (see McDougall 1997: 189).

The Act itself (available on the history.nasa.gov website) begins with the proposition that 'it is the policy of the United States that activities in space should be devoted to peaceful purposes for the benefit of all mankind'. It then continues that national considerations of 'general welfare and security' demand that adequate provision be made for aeronautical and space activities, and that activities to this end should be the responsibility of a civilian agency, excepting such activities associated with weapons systems, military operations or defence, which shall fall to the Department of Defense. It proceeds to define the objectives of such peaceful 'aeronautical and space activities', to wit, the expansion of human knowledge of the atmosphere and space, the development, safety, speed and performance of appropriate vehicles, and the creation of vehicles for research purposes in space, 'capable of carrying instruments, equipment, supplies and living organisms'. It also adds a series of more general objectives: setting up 'long-range studies' of the benefits and problems to be anticipated from such activities dedicated to 'peaceful and scientific purposes', preserving the role of the United States 'as a leader' in space science and technology and its application to peaceful ends 'within and outside the atmosphere', co-operation and exchange of research with relevant defence institutions and, likewise, cooperation and exchange of information with 'scientific and engineering resources of the United States', to avoid 'duplication of effort, facilities and equipment'.

The rest of the Act dealt with matters of governance and the ownership of patents created through NASA sponsored research. Both are crucial elements in the relation of the state to civil society, to the new agency in the first case and to corporations and universities in the second.

Extra-terrestrial communication

As McDougall comments, the Act left ambiguous the relation between civilian and military research; while it moved towards the 'state-directed mobilization of space and technology', it did not commit the nation to an all-out space race; and though it mentioned several goals for research and development – prestige and scientific advance among them – it gave no order of priority between them (see McDougall 1997: 172). Underlying all these ambiguities there was the basic fact that 'almost all space technology could be put to military use as well as civilian use with no way of sorting it out' (McDougall 1997: 174). A straightforward example of this overlap, 'precise measurement of the shape of the earth and its gravitational and magnetic fields was a prerequisite to improved missile accuracy' (McDougall 1997: 191). In such a world, technical and scientific progress cannot be separated from the immediate strategic context, when any advance may give a temporary advantage or, on the contrary, may represent the recovery of a perceived or real lag behind a rival innovation. It was the political and military context which determined significance, as the example of Sputnik makes clear; the feedback between technology and its context of interpretation was tightly drawn and continuous. The analysis supports McDougall's thesis that the space programme in its original phase was coterminous with the Cold War; technology signalled political concerns at every level – a case study of Kittler's thesis (1987) linking military technology with both communication and concealment.

By the end of 1959, there was a balance to be struck between mutual deterrence and the pursuit of 'stability' through arms control, with the proviso that a missile test ban would inhibit the space programme, which was the principal way forward to ending Soviet secrecy and so verifying any arms control agreement (McDougall 1997: 193).

NASA's early development took place in this context, with a following wind in terms of national politics but with serious potential rivals in the military sphere. NASA inherited NACA's existing facilities – research laboratories (at Langley in Virginia, Ames in California, and the Lewis flight propulsion laboratory), a test station for high-speed flight (at Edwards AFB) and a rocket test range (at Wallop's Island). It took over projects and sites from the Navy and ARPA and the Army's big rocket engine programme, and developed a new centre, the Goddard Space Flight Centre in Maryland (see McDougall 1997: 196–197).

It is in the interface between the agency's public aspect – its focus on internationalism, the public good, science and public education, played out in front of an audience, on television for the most part – and the necessarily secret aspect of surveillance, defence and the development of new weapons, that we should place the development of projects concerning extra-terrestrial communication. Such projects were absent from the agency's stated aims at legislation (above), but nevertheless gained plausibility from the direction of travel we have been tracing. We can point to several elements we have met that could readily be extended to include the possibility of life elsewhere, elements which lent themselves to drawing attention away from the needful business of surveillance and preparing for war and raising the public's horizon to new ideas of progress and hope. This extension is a matter not of influence but of congruence: once focussed on the potential of communication and exchange, these are narrative elements which can come along without resistance.

The first is the idea of 'reaching beyond this planet', from a statement, for example, made to a United Nations' meeting held in 1958 (McDougall 1997: 184). The second concerns discussion of an internationally agreed legal framework allowing the exploration of space (McDougall 1997: 188–189), which contains – only implicitly – the potential for further extension beyond this planet, implying the possibility of interplanetary agreements. Likewise, the theme of the protection of satellites and defence from attacks from space (McDougall 1997: 191–192) can have implications of interplanetary threat. And the wider themes of international prestige, how to relate to emerging nations, and the re-imagining of relations with the poor at home, could all be taken over without remainder into the possibility of encounters with other civilizations. Indeed, this trio of ideas, prestige, cultural contact, and the reversal of accepted power relations, has been present in the thought of interplanetary contact ever since Wells' *The War of the Worlds*. It is not that scientific ideas and technological inventions become captured by speculation of an unscientific kind, but, on the contrary, that scientific research and the accompanying political context lend reality to ideas transcribed into fiction from an earlier situation.

But apart from such matters of atmosphere, there are also positive examples of research projects aiming at communication, receiving and sending messages.

III. SETI in NASA

The idea of listening for signals originating from other planets is exactly contemporary to the creation of NASA. Such a project necessarily implies technologically advanced civilizations elsewhere, capable of sending signals and therefore of communicating. The idea of searching for signals from space took clear form with the publication of Cocconi and Morrison's 'Searching for interstellar communications', published in *Nature* in 1959. This paper established 'the radio region of the electromagnetic spectrum as a logical place to search for signals from extraterrestrials' (Billingham 2014: 38). Billingham, who was the officer above all others responsible for establishing such research at NASA, also identifies two other inaugural events in the field, both associated with the name Frank Drake. Drake, based at the National Radio Astronomy Observatory in

Green Bank, West Virginia, conducted the first search listening for possible intentionally-formed radio signals, Project Ozma, in 1960, and, the following year, convened a small meeting of interested scientists at Green Bank, sponsored by the National Academy of Sciences Space Science Board, with these objectives: 'to examine the prospects for the existence of other societies in the Galaxy with whom communication might be possible, to attempt an estimate of their number, to consider some of the technical problems involved in the establishment of communication, and to examine ways in which our understanding of the problem might be improved' (cited by Billingham 2014: 39).

We should make two observations. First, this trio of events comes within the orbit of orthodox science and state funded research; the names of *Nature*, the National Radio Astronomy Observatory and the NAS Space Sciences Board are markers of this insider status; the ideas and practices involved were not eccentric or idiosyncratic, however innovative they might have been. Second, this 'origin' for the contemporary search for extra-terrestrial intelligence is deeply conventional; every account begins from Cocconi and Morrison, the Ozma Project, the Green Bank workshop and, among its results, the 'Drake equation' for calculating the number of potential civilizations in the galaxy from whom we might receive signals. This agreed account has two reasons. On the one hand, we are dealing in a small group of overlapping interested parties – researchers, laboratories, institutions, and funders – and, on the other hand, this group also provide their own historians. Because of its concern with public education and shaping wider concerns, NASA has been aware of the need to document, record and present an account of the history of each project of space exploration and every turning. Many reviews are by the actors or by the historians employed for the task; they are all participants in a common project and there is a realistic ethic of acknowledging the importance of the team. As a result, the sense is of a cumulative project of learning, there is little consideration of breaks or changes in presuppositions, and outside or wider factors rarely come into the story, except, for example, when political decisions not to support funding impinge. In short, the story is told from the project's present, and the past introduced in terms of its progressive elements, selected from the perspective of present achievements. We rarely consider parallel cases, alternative genealogies, wider repeating patterns, or such matters as shifts in the background categories that close and open possibilities, let alone introduce counter-factual speculation, as McDougall does when he says: 'we are still us[ing] updated versions of 1940s German technology. In the long run, the chemical rocket is just not the key to the future ...' (McDougall 1997: xvii).

All three events – publication, the first search, and workshop – were in fact conceived independently of NASA's concerns, and questions of life elsewhere and its intelligence were marginal to the challenges of rocketry, engineering and planetary exploration confronted at the outset. Nevertheless, NASA had interests in 'exobiology'[15] from 1960, looking for signs of microbial life beyond the earth in the margins of projects of taking samples from the moon, meteors and other planets, building on earlier work such as theorems of the chemical evolution of life and the laboratory synthesis of amino acids.[16] Recommendations to NASA from the NAS Space Science Board in the 1960s made regular reference to the search for signs of extra-terrestrial life among other research aims. At the same time, events abroad, a Russian search for signals in 1963, a 'conference on extraterrestrial civilizations' the following year, also organized by Russians, a proposal to establish an international symposium on Communication with Extra-terrestrial Intelligence (CETI) in 1965, and the English translation of Shklovsky's *Intelligent Life in the Universe* (Shklovsky and Sagan 1966), all mirrored the concerns reflected in the initial research impulse (see Billingham 2014: 39).

Exobiology and monitoring for extra-terrestrial signals separate around the single question of the evolution of intelligence; with hindsight, there is an enormous intellectual distance between taking samples looking for traces of the components of organic molecules and searching for signs of an advanced planetary civilization. NASA only formally engaged with the search for extra-terrestrial intelligence in 1969 (Billingham 2014: 40f.), the year that the Air Force achieved its quittance from any engagement with

15   A term credited to Joshua Lederberg – see Lederberg (1960).
16   See Billings (2012: 3); this article maps the three-way distinction between exobiology, astrobiology and SETI.

flying saucer investigations. Exobiology and what would become known as SETI were distinct projects, both housed in the Ames Research Center.

Under Billingham's administration, an initial feasibility study in 1970 led to a design called Project Cyclops in 1971. The project proposed an array of radio telescopes and signal processing to search for incoming signals in what was thought to be the optimal part of the microwave spectrum, focussing around the spectral lines of hydrogen (the commonest element in the universe) and the hydroxyl radical, for it was assumed that water would be essential to the development of 'Extraterrestrial Intelligent Life'. This project remained on the drawing board. Other proposals followed, and in 1974 the NASA Office gave its first (small) funding for searching for interstellar communication (Billingham 2014: 44). In Billingham's account, despite encouragement to pursue such projects in such a formal document as the National Research Council's decennial report on astronomy and astrophysics (a product of its Astronomy Survey Committee in 1972–1973), interstellar communication remained 'outside the respectable norms adhered to by most of the scientific community' (Billingham 2014: 45). As a director of research at Ames, he continued to promote the idea and put together a series of workshops on SETI in 1975 and 1976 in which, as he notes, interests in discerning intentional signals overlapped with two new topics, extrasolar planetary detection by astronomers, on the one hand, and an interest among exobiologists in the evolution of intelligent species and technological civilizations, on the other (Billingham 2014: 46). At least the first of these was a sign of developments leading to the transformation of exobiology into what would become known as 'astrobiology'. Billingham notes that the name SETI (rather than Communication with Extra-Terrestrial Intelligence) was adopted in December 1975.

The report of the workshops confirmed the microwave window as a place to begin research and noted great progress in the technology permitting spectrum analysis in terms of speed of handling and volume of data. Billingham became Chief of the Exobiology Division at Ames, where he constituted a SETI Program Office in 1976. Under his leadership, a bimodal research strategy was proposed: a focussed radio telescope beam, continuously monitoring selected target stars, achieving high sensitivity (on the model of Project Cyclops), and a broad survey, sweeping the beam across the sky, seeking total coverage with reduced sensitivity. The programme entered into an agreement with the Jet Propulsion Laboratory (JPL), Ames to pursue the targeted search using existing large telescopes around the world as antennae, and the JPL to use its telescopes at Goldstone in the Mojave Desert for the broad survey.

Under Billingham's leadership, there was steady progress in terms of research, credibility, and institutional embedding. Ames, JPL and NASA formed the SETI Science Working Group in 1980, which produced a report in 1984 in which it 'confirmed the microwave region as preferable; endorsed the bimodal strategy; and envisaged a five-year R&D effort to design, develop, and test prototype instrumentation'. Its first conclusion was: 'The discovery of other civilizations would be among the most important achievements of humanity' (Billingham 2014: 49–50).

Despite the small sums involved (around 0.1 per cent of NASA's budget – Billingham 2014: 52), the project was singled out for criticism by a politician, Senator William Proxmire, who, in 1982, introduced an amendment to the NASA budget to eliminate all funding for SETI; this element was however reinstated in 1983, for key administrators and scientists put in work to lobby and the Senator listened. Despite the warning signal, SETI flourished at NASA for the next decade. At the same time as Proxmire's cut in 1982, the Decennial report of the Astronomy Survey Committee supported SETI research. The proposal for five years' R&D funding was accepted by NASA and, between 1983 and 1987, the project, now called the Microwave Observing Project, received around $1.5 million a year for research, leading to the development of digital technology for new spectrometers (Billingham 2014: 55). A new SETI plan was proposed in 1987, putting forward a ten-year search for narrowband signals, both targeted and 'sky', with a cost of $73.5 million (Billingham 2014: 56). NASA agreed in 1988 to initiate the project; funding at this stage was running at $3 million a year (Billingham 2014: 60). SETI took on the status of an 'approved NASA project' in 1990, and the final development and operations phase was begun. The budget for 1990 was $6 million, rising to $16.8 million in 1991. The final project plan, in Billingham's account, 'outlined a 10-year search at a total cost of $108 million' (Billingham 2014: 60–61); the project had 140 people working on it and was inaugurated on 12 October 1992.

The name of the project was changed to the High-Resolution Microwave Survey (HRMS) in the same year, and its 1992 budget was $17.5 million. The signal detection system was shipped to the Arecibo Telescope and the sky survey instrumentation tested at Goldstone, and searches carried out, generating enormous flows of data, for the next year, reported on in August 1993.[17]

However, at this point, funding was cut. An amendment was introduced and passed in the Senate removing the HRMS from the 1993 NASA budget. Grants and contracts had to be closed down and the teams dispersed.

SETI reclassified

SETI ceased at this point to be part of the state-funded civilian space programme. The story bears certain parallels with the side-lining of the Air Force's Project Blue Book in 1953. In each case, a minority view carefully nurtured a prospect of new research, working in a complex organizational environment, and succeeded first in gaining a foothold and then, after initial feasibility studies and the development of technical means, gained sufficient traction and financial support to undertake substantial investigations, only to be met in quick order by opposition from outside sources which put an end to the project. In each case, too, removing official support for the project simply displaced its main sites of operation.

By 1993, SETI was not confined to NASA; since Drake's initial survey, other searches had been conducted both in the United States and abroad; a variety of interests, including the question of maintaining appropriate standards in the science and engineering, had been pursued through a range of international organizations, conferences, and publications; and there had been discussion through international bodies of the legal and prudential forms appropriate if a signal were received, including international declarations of principles, whether to reply to any signal detected, what might be conveyed in such a reply, and the appropriate decision-making

---

17 Billingham's article was originally published in 2000; the same ground is covered in more detail in Dick (1993). For a more recent account, see Scoles (2017).

processes. NASA contributed to all these discussions (see Billingham 2014: 57–60). So, when the NASA survey was closed down, SETI continued by other means. In particular, the targeted search was taken over by the SETI Institute, founded in California in 1983 (Billingham 2014: 54), which ran Project Phoenix from 1994 to 2004, and the SETI League, formed in 1994, began a new all-sky survey in 1995 using Project Argus, 'a wide-sky, broad-frequency, low sensitivity search with small telescopes' (Billingham 2014: 57). Nevertheless, 'although HRMS was a very small project by NASA standards, it [had] dwarfed all other SETI efforts combined' (Garber 2014: 93); SETI projects now had to rely on raising funds from private donors, state university sources and a variety of national science grants. SETI had crossed a boundary which had been much debated in the Eisenhower years and earlier; it was no longer formally part of direct state support for technological development and scientific research. In practice, the boundary was less clear; NASA loaned equipment to Project Phoenix, for example, and personnel continued to move between NASA, university departments and non-profit funded institutions.[18]

Can we read any significance in this act of reclassification? Garber (2014, originally 1999) attributes the closure of the programme to a range of contingent factors: 'anxiety over the federal budget deficit, lack of support from some sectors of the scientific and aerospace communities, and unfounded but persistent claims that linked SETI with non-scientific elements all made the program an easy target in the autumn of 1993' (Garber 2014: 70–71). Yet none of the elements was unique to the moment. The budget savings were in any case small; there are always contradictory scientific voices to be called on by critics, and it can be claimed that, in the face of any budget cut, that more could have been done by administrators and leading scientific advocates. Likewise, the potential to associate SETI with science fiction stories has always been present. But there were other redistributions of values specific to the time.

The first was the end of the Cold War. The Soviet Union had ceased to exist and the entire justification for the space programme had had to be re-thought. This re-thinking was not the matter of a moment, but the

18 For a detailed account of SETI post-1993, see Scoles (2017).

immediacy – and potential threat – of imagining and preparing for contact with another civilization took on new contours. Carl Sagan's deeply informed novel about contact from this period, drawing on his experience of the SETI project, is striking for the benignity of its images, its internationalism, and its belief that scientific interest alone is sufficient motivation for a long-term, state-funded project (Sagan 1985, discussed below).

The other is the more recent re-conception of the justification for space exploration in terms of what became known as astrobiology (as distinct from the interests of physical cosmology and – always in the background – on-going defence needs). This offers a case where shifts in scientific presuppositions and innovations in what constitutes evidence may be held to have had effects in the wider cultural setting. I will offer a discussion from this perspective, not, then of the scientific basis as such, but of the narrative elements that can be supported on this basis.

IV. Astrobiology

With the end of the Cold War, the rationale for the Space Programme needed to alter in form, with the covering arguments for the peaceful exploration of space taking on more substance and bearing weight as the military justification lessened. In this context, one compelling aspect of the argument for SETI, anticipating the possibility of extra-terrestrial competitors mirroring the real threats posed by human rivals, retreated, but the underlying supposition of such space exploration – that we might encounter other minds interested in our kind of life and prepared to assist our development – persisted and could take on new forms. In short, while astrobiology represents a straightforward shift in scientific presuppositions, based in developments in technology and biological ideas, certain features of this shift lend themselves to retelling an older narrative.

There are two features in particular that are of interest in this perspective. On the one hand, the sense of a continuity between all material forms, from the largest scale to the smallest, so that everything can be assembled into a single developmental narrative and so also into a single timeline. This assemblage can be recruited to support an assumption of direction or purpose – a teleology – including a narrative concerning the coming to self-consciousness of the Universe in contemporary human scientific understanding. On the other hand, in addition to the mechanisms of variation and selection, the new techniques employed point to developments in the existing explanatory schemata, incorporating contingency and self-organization in new ways. Both in the principle of continuity between scales and the identification of pressure points in the explanations of order, these new scientific forms present formal resemblances to the theosophically inspired narrative[19] concerned with the possibility of extra-terrestrial communication.

We might, in short, put a question: why is it plausible for experimental scientists – biologists, cosmologists, engineers – to entertain the possibility that there may be a single process (in broad outline) of the development of life in the universe, including the chances of the repeated creation of intelligent life and of the evolution of technological civilizations? We can, it appears, assume there will be points of precipitation when the order of complexity increases, points which correspond to our own overall planetary self-understanding. We are tracing the continuation of some long-standing narrative themes which serve as uncontested presuppositions and, even if contested in detail, nevertheless set the frame. Once we have glimpsed this frame in the materials, we can ask, in the next chapter, what part does the idea of 'communication' play in supporting these presuppositions?

Astrobiology – narrative continuities

One aspect of astrobiology is then that it offers an account that might join all material forms, inorganic and organic, into a single continuous narrative, from the formation of the Universe to our present civilization, listening for other voices.

---

19   Considered in the second essay, *Religion and Science Fiction*.

This evolutionary narrative has been present in potential since Darwin's application to living forms of law-like processes worked out on inorganic systems, substituting questions of variation and selection for those of design and vitalism. This account has been further developed, first in the 1920s with the application of statistical models derived from thermodynamics to populations considered as distributions of genetic variation, and, more recently (since the 1980s), with developments in the study of dynamic complex systems, with computer-aided mathematical models dealing in the structural transformations of large populations showing random activity (see Depew and Weber 1995). In each case, models deriving from consideration of the dynamics of inorganic systems (from physics) have been applied to organic systems, showing how new stable states may emerge from random change, in the form of the possible origin of life, selection of individual variation, the separation of species, and the formation of the higher orders, the broad division of living things into orders, classes, and kingdoms.

The potential is therefore there before the advent of astrobiology which, however, emphasizes the cosmological context by projecting questions of the origin and evolution of life from Earth to other planets. This is possible because the common models allow a single story, with biological evolution placed within a wider account of cosmological origins and development (see Morowitz and Smith 2016).

That is the interest of the Drake equation, which early on assembled the elements that are needed to join the evolution of the Universe, the appearance of life on other planets, and the idea of extra-terrestrial civilizations with technology capable of generating radio signals which can be detected on Earth. Its aim was to estimate the number of advanced civilizations in the Milky Way whose electromagnetic emissions were detectable and with which humans might communicate. This number (N) relates to (a) the mean rate of star formation, (b) the fraction of stars that have planetary systems, (c) the mean number of planets in each solar system capable of supporting life, (d) the fraction of potentially life-supporting planets that actually develop life, (e) the fraction of planets with life where life develops intelligence, (f) the fraction of intelligent civilizations that develop a technology that releases detectable signs of existence, and (g) the mean length of time that such civilizations can communicate.[20]

Subsequent research and technical advances have given more substance to several of these elements. For example, the NASA Kepler satellite programme has increased the number of known stars with exoplanets and the proportion of those exoplanets with inferred 'habitable zones'. In parallel, changes in understanding of the earliest ancestral forms of life on this planet and the conditions under which they arose have increased the range of planets which might be supposed to be life supporting. In this fashion, the bar has (in theory) moved towards the possibility of other civilizations capable of communication.[21] A contributing factor has been new ways of thinking about the classification of living things based on handling large quantities of data, which point to the early importance of systems which took their energy from their surroundings to the evolution of life on earth. Astrobiology is marked by these three conditioning discoveries: we are becoming aware of large numbers of potential earths at the same time as we have broadened the notion of the zones in which life might emerge and the kind of life forms we might find operating within these zones (see Domagal-Goldman and Wright 2016).

We might remark how astrobiology integrates the concerns of inorganic and organic life. Neither the equipment used – spectroscopy and the potential identification of both 'biosignatures' and inorganic compounds – nor the techniques of analysis – handling large quantities of data – make any distinction between the biological and the physical or chemical, and indeed the 'building blocks' of life comprehend both organic and inorganic processes. We are looking at some kind of integration between cosmology and the nature of life, which some might interpret as signalling an application of physics techniques to biological problems (see Dick 1998).

For this reason, introductions to the possibility of extra-terrestrial life from the more recent period (e.g. Heidmann 1995; Grinspoon 2004)

---

20 Taken from the Space.com website.
21 See, for example, 'Are we alone in the universe? Revisiting the Drake equation', NASA Exoplanet Exploration, 16 May 2016.

draw on a narrative which, at least on the surface, resembles Drake's. Cosmic evolution has been integrated into a series of stages, passing through the formation of matter, of gases, of stars, of planets, of organic life, and of intelligence. And, in part because of the advances in techniques, understanding, and detail, we may have confidence in ourselves as models of this intelligence. Indeed, the cosmos is becoming aware of itself in the present work of scientists here on Earth; hence Grinspoon's claim that 'the universe has grown up to an age where it wants some answers about its own provenance. Here on Earth, the cosmos has awakened from a 12-billion-year dream. It seems that our consciousness, in inchoate form, was here all along, waiting for the right conditions to precipitate out of inanimate matter'. To which he can add, 'Elsewhere, is it slumbering still, or were we among the late sleepers?' (Grinspoon 2004: 87).

Astrobiology then not only integrates biology and the material sciences but also has the potential to include the emergence of consciousness within research of the nature of matter. It is open to us to concern ourselves with whether the kind of intelligence we display – or the consciousness of which we are at the threshold – has been anticipated in the formation of more advanced civilizations elsewhere, particularly while those possible others remain silent. In short, astrobiology continues to allow for the search for extra-terrestrial intelligent life; an advanced cosmic consciousness may be located in ourselves or found elsewhere.

Early in this history, a biologist, Ernest Mayr, criticized what he saw as an unstated assumption behind the Drake equation. He rejected the possibility of intelligent life evolving twice, and identified an implicit determinism in the arguments of physical scientists: the physicist was

> certain ... that if life originated on Mars (or had been transported to Mars), this would inevitably lead to intelligent humanoids. The production of man was for him like the end product of a chemical reaction chain where the end product can be predicted once you know with what chemicals you have started. He took it virtually for granted that if there was life on a planet it would in due time give rise to intelligent life ... [The argument runs] if organic evolution on earth culminated in intelligence, why should it not have resulted in intelligence on all planets on which life had originated? (Mayr 1985: 25)

Mayr's argument[22] contains a kernel of truth; statistical arguments of this kind become determinist when extended over very large numbers; every possibility, however improbable, may be fulfilled. The argument also contains a second unexamined presupposition, that the evolution of intelligence is in some sense the justification of the whole process. The other way about, Mayr suggests, 'an evolutionist is impressed by the incredible improbability of intelligent life ever having evolved, even on earth'.

It is worth pointing out how much of the standard theosophical cosmology has been replicated in this picture that Mayr rejects. The history of the universe has been drawn as a series of extraordinarily complex cycles of development with, as a defining moment, the emergence of the self-consciousness of the entire system in the present human moment, here on Earth. As a bonus, this moment may have been anticipated many times before, elsewhere in the universe, so we may look for communication from more advanced civilizations, perhaps learning from them as we negotiate our dangerous times and adapt. The intelligent minds from elsewhere need not be given the central role, but they could take it on.

Astrobiology – new explanatory schemata

Astrobiology shares in wider developments in the concerns of contemporary biology. Rather than relying on the idea of natural selection as the principal characteristic of life, two other features, always present in the story but pushed to the sides in earlier versions, have come to prominence: the operations of chance and the powers of self-organization of systems. Once we deal in the properties of large systems composed of random events, recent work points, on the one hand, to the importance of initial differences in 'final' outcomes, due to the stabilizing properties of large systems and, on the other, their capacity to undergo shifts to different 'phases' or orders of complexity, from one kind of stable state to

---

22 He was revisiting a debate that had emerged in 1965 and he chose to identify a particularly dogmatic opponent.

another. The question is how – if at all – do these properties lend themselves to narratives of life elsewhere?

The repositioning is largely a matter of the scale being considered. By focussing on selection at the level of the individual organism, Darwin supported a gradualist account of the development of species and evolution of kinds in lines of branching descent and offered no strong explanation for the existence of the higher taxa. This account was coupled with the absence of any mechanism to explain heredity and the lack of evidence for natural (as opposed to artificial) selection, so that a variety of developmentalist accounts returned to the fore, more or less eclipsing the Darwinian theory of natural selection by the end of the nineteenth century (Bowler 1983). The emphasis shifted in the 1920s with evolutionary theory being fused with the emerging science of genetics, producing something called the 'Modern Synthesis' (Julian Huxley 1942). This fusion was created not based on the natural law model of the original theory but by recourse to the statistics of populations, applied to the totality of genes in a breeding system. This statistical model had been developed in the late nineteenth century by physicists to save the Newtonian model from exceptions emerging from atomic and molecular theory; biologists subsequently saw that genes could be treated in the same fashion as atoms or molecules. This approach not only allowed the theory of selection and adaptation to be rethought, but also provided a convincing account of the evolution of distinct populations, or speciation.

The Modern Synthesis also had its limits: it came to be challenged by developments at the small scale, in accounting for some molecular and cellular processes, and at the large scale, as before, of explaining the higher order separations of living things into orders, classes and kingdoms. At either scale, it is not clear that natural selection alone can explain evolutionary processes; other factors, notably the operations of chance and the self-organizing properties of biological systems, come into play. The Modern Synthesis has then in turn been supplemented since the 1980s by developments in the study of the dynamics of complex systems, in terms of new mathematical models which, on the one hand, look at the emergence of stabilizing properties in the behaviour of large populations of objects moving at random, and on the other hand, describe transformations in the overall state of these populations, whereby they can move from one stable state to another. In short, we are getting descriptions of both the formation of boundary conditions and of phase shifts as features of large populations of random – or chance – activity. These dynamic models allow operations of chance, selection, and self-organization to be integrated in a new fashion. The recent adoption of non-linear statistical techniques offers new ways of thinking about a series of topics – the origins of life, the development of the organism, the organization and classification of types of life forms, and ecology – which have never fully been incorporated within the logic of natural selection nor within the Modern Synthesis (see Depew and Weber 1995).

These repeated innovations in terms of scale bear a formal resemblance to the conditions under which the original theosophical synthesis emerged, which used the appearance of new theories as evidence for the need for covering narratives involving intentional actors. Just as the recognized processes of Newtonian physics became confined in the early twentieth century to a middle ground, applicable neither at the very large nor at the very small scale, so in turn, under the terms of the most recent biology, natural selection appears to be confined in its application only to a middle-order range of circumstances. This confinement of 'orthodox' explanation to an intermediate zone formally resembles Madame Blavatsky's intuition that other principles operate at other scales and, likewise, that this shift in focus complements the emergence of mind. Despite their different statuses, there are resonances between the two cosmologies.

Contemporary shifts in explanatory models support two possible outcomes to the question of the emergence of intelligence within cosmic evolution; they are, in brief, the potential for contact in the form of communication, or, alternatively, silence. For if you take the phenomenon of 'self-organization' – at all kinds of levels, from the innermost workings of the cell to the biosphere (and beyond) – you are returned to earlier concerns of 'developmentalism': form-focussed thinking which offers hope for the possibility of life 'out there'. Forms which, while based on different organic principles, may show correspondences with terrestrial life forms and even (perhaps) in some instances with human concerns, such as the desire for communication, ordered social life, and the development of technologies and civilizations. In essence, the emergence of intelligence is reconceived. These questions of 'convergence' underwrite not only SETI but exobiology and astrobiology more generally: they are interesting questions, potentially threatening, and they resemble some of the speculations made by science fiction writing. The point is, once you have formal mathematical models which reintroduce these questions, even speculations return on the new basis. Although SETI has been on the backburner for a decade or so (as far as NASA is concerned) and talk of other civilizations has muted in favour of exobiology and now astrobiology, if we take seriously the implications of the kind of statistical models being exploited, these larger questions of convergent developments and direction of travel will also emerge.

However, if you turn to the question of chance and its return to prominence in this newest synthesis, with both the role of initial conditions in the subsequent stabilization of distinct complex systems and their potential phase shifts to higher levels of complexity, which may stabilize in their turn, there are at least arguments that life on Earth is due to so improbable a sequence of events – from the appearance of the basic building blocks for cells (lipids, amino acids etc.) to the self-sustaining biosphere which prolongs life and its possibilities – that, no matter how many exoplanets we locate, and no matter how wide we expand the 'life zone' and the life forms that may arise within it, we will never get to more than a single instance – life on earth, and that life exceedingly fragile. This is a version of Mayr's scepticism. We may never escape from the unique miracle of life on Earth, alone in the Universe of infinite stars and planets. So, while astrobiology seems with one hand to offer extraordinary possibilities of life elsewhere, considered as a serious possibility for the first time, with the other hand it takes away that possibility, for there are statistical questions of the sheer improbability of finding life elsewhere, no matter how many exoplanets 'within the life zone' are discovered by ever-more powerful telescopes. There are so many extraordinary chances needed at so many levels of organization that we may never avoid the singularity of life on Earth.

In sum, even on new grounds, the alternatives appear to be communication or silence. To return to the question of the closing down of the SETI programme, in the period of the emergence of this synthesis – NASA formed an astrobiology programme only in 1997 – with the disappearance of any rival space programme and increasing confidence in human scientific consciousness as the coming to self-awareness of the entire process, there was an opportunity to shed the search for extra-terrestrial intelligence. On the one hand, such a possibility was fully integrated within the theoretical system as a possible item, but, on the other, it could be seen as being simply 'curiosity led' research (see Garber 2014: 77) and as such not an appropriate recipient for directly voted programme funding. What remained was a framework constructed around the potential for contact and exchange, of friendship or hostility. It is this framework which we now need to consider.

CHAPTER 2

Communication and its failures

We have considered two instances – the Air Force ridding itself of any concern for UFO sightings, on the one hand, and, on the other, the contemporary development of a civilian-led space programme which first took on seeking for signs of other, extra-terrestrial civilizations and then marginalized that concern while, at the same time, bringing the evolution of consciousness within a cosmic framework to constitute a central puzzle. If there is a category that holds these two independent histories together, it is that of 'communication'. In the first case, the possibility that something was being communicated by the appearance of craft of unknown provenance, followed by realization that these hopes were interfering with practical arrangements for communication in time of crisis and efforts had to be made to quieten the interest aroused and even to re-educate parts of the population. In the second case, dealing in quite different materials, a similar pattern: hopes of disembodied contact, dashed in time by practical considerations backed by the delay in finding any substantiated signal. In both cases of suppression, the hope of communication was not extinguished but took other forms.

I. Two senses of communication

'Communication' has become a term which includes every aspect of life, whether biological, human, or social, as if it were a key to effective functioning at every scale and, correspondingly, its failure the explanation of things going wrong. Communication in this sense can be broken down into the reading, storage and transmission of what we now call 'information', and 'right' reading, storage, and transmission can be regarded as the basic requirements for the operation of cells, organisms, and populations, and, in the human realm, as needed for the proper operation of minds, relationships and every kind of collective activity, whether technical, economic, or political.

Yet the term demands reflection, for it is a recent form: to quote one source, 'only since the late nineteenth century have we defined ourselves in terms of our ability to *communicate* with one another' (Peters 1999: 1). This conclusion is drawn from a recent history of the idea of communication, John Durham Peters' *Speaking into the Air*, which I shall draw on to help us understand the moral coding with which we are concerned. Peters' thesis has a number of aspects; we can draw three lessons from his account.

In the first place, he identifies two senses of the term. In earlier usages, 'communication' signified 'making things in common' or 'joining together', with no sense of exchange or dialogue. It concerned assemblages of persons and communities such as acts of distribution and public rites, offerings and participation, not, primarily, 'a message-sending activity' (Peters 1999: 7). Think of the *Prayer Book* term 'to communicate', meaning to take communion, to share in the Body of Christ, to participate in the risen life of the community; it does not evoke the idea of exchanges of information.

In its more recent and now dominant form, however, 'communication' emphasizes the idea of individuals making contact through speech or signs; 'communication' in this second sense signifies a transparent reciprocity of meaning between minds, the sending and receiving of a message, a private exchange of ideas without distortion. Distortion can then be attributed, in this second sense of the term, to social forms such as rhetoric or tradition: the original meaning is forcibly rejected.

We will shortly look at the genealogy of the second form and some of its implications, both the opportunities it offers and its limits. But first we ought to ask, how do we evaluate these alternative forms? We might put the contrast in these terms: humans form projects together of a loose kind – for example, the formation of a family and bringing up children, or running a business, or teaching and learning – and language is one of the tools they use in carrying out these projects. But to imagine the exchange of accurate information between minds is the defining feature of the project is an illusion. Frequently, the different actors will have distinct and at points incompatible accounts of their common aims, but this does not prevent successful cooperation. Cooperation is a better term for understanding sociologically what is going on than construing the processes in terms of better or worse mind-to-mind contact, or communication taken in its second, modern sense. The distinction may be thought of as a matter of scale; humans interact and create their world on a limited range of scales through embodied means, including the use of language. Turning to the notion of 'communication', however, tends to focus our attention either above that human range or below it, for communication is conceived as dealing in disembodied information, and as therefore occurring either at too large or too small a scale for humans to participate effectively. The communication of information becomes detached from any human context and appears to take place elsewhere, either in secret places, between people who could be conceived of as conspirators, or else as tuning into the secrets of the Cosmic Mind, exemplifying the largest scale.

In Peters' account, the first, older sense is more reliable than the second, more recent. And if we place weight on the second, he suggests, we meet this modern sense of communication not through moments of transparency, which are, alas, too rare, but through its failures, when signs go astray or do not arrive. The reason for this is that 'signs are fundamentally public, that is, capable of multiple junctions of meaning' (Peters 1999: 168 – a summary of Peters' thesis), and so are always 'open to eavesdropping' and to being broadcast. Failures to communicate arise in practice in common encounters involving actual people, they are situated and bound in particular contexts, and they are not because minds have failed to meet; in Peters' formulation, failures of communication are 'erotic', not 'noetic'; they involve persons – bodies – and not simply minds.

Yet, the second idea predominates in many contexts and thereby generates its own insoluble problems. For example, the search for extra-terrestrial intelligence is constructed around failures to communicate. You may hope for messages from newly detected craft displaying evidence not only of advanced technology but also of intelligence and intention, but what you get are wrongly focussed popular expectations which have to be met with propaganda and attempts at re-education on the part of the authorities. Likewise, you may search for messages from other civilizations, but none arrive and, if they did, their meanings would be inscrutable and the senders long dead. In these accounts, every hope turns on contact between minds and is frustrated by bodies, whether present or absent.

Peters places the particularity of communication, then, not in the 'interior sovereignty of the mind to arrange things at will', but in the specific arrangements of persons that constitute the setting of human behaviour, and he proposes that the 'pathos' of failures to communicate 'is not that minds fail to share the meanings of signs but that mortal beings miss getting in touch' (Peters 1999: 168). By and large, modern notions of 'contact' eliminate any consideration of mortal bodies and focus on imagined disembodied minds, as we shall see. Because of this act of repression, they are haunted by the spirits of the dead and their ilk.

A genealogy of the modern sense of communication

In the second place, where does this modern sense of 'communication' come from? Why do we think of communication as the direct transfer of thought from one mind to another? Peters offers us a pre-history, tracing elements which make contact over time, forming layers of the past. He begins from Socrates' distrust of writing as a tool for sophistry, in contrast to the directness and sincerity of speech. He then touches on Augustine's seeking after a concept of interiority in prayer. But the turning point is Locke's theory of the sign, for which mention of 'the interior sovereignty of the mind' (above) might have prepared us. This theory has a place in any history of the development of Protestant thought. For Locke, the sign is a publicly oriented counterweight to the inwardness of the self and its possession of meaning; the mind having become enclosed, and thought being its property, the problem becomes how men form common projects and share their thoughts between minds, and the solution, that they exchange signs. And there appear to be two traditions, closely intertwined, which respond to these two constraints of inwardness and sharing: in the first place, the dream of a purified scientific language, which would be transparent, a passive tool to convey univocal meanings, and, in the second, the dream of direct, wordless communication between minds, developed through Mesmerism, Spiritualism, Theosophy and beyond. The project of a scientific language and metaphysical accounts share a common parentage.

In practice, it is this double inheritance and the interaction between its components that gives us the modern concept of communication. Peters traces this history in some detail. In essence, each scientific discovery simultaneously is translated into a technology that allows messages to pass and is conceived in terms that talk of the transfer between human minds. The practical starting point was Newton's discovery of gravity and the forces passing between heavenly bodies, which was understood by Mesmer a century later to apply also to the human sphere in terms of influence, healing, and the adjustment of relations. In this fashion, material forces were reconceived in moral terms.

Locke is the precursor, then, not so much because of his development of possessive individualism, but rather because, to shape his account of mental powers, he resorted to models drawn from the seventeenth-century physics of bodies. The metaphor of action at a distance was crucial. 'The concept of communication as we know it originates from an application of physical processes such as magnetism, convection, and gravitation to occurrences between minds' (Peters 1999: 78). These processes offer instances of influence passing between separate bodies[1] and so could serve to illustrate the task of overcoming the isolation of minds. New conceptions of the order of matter serve to shape a new understanding of mutual relations between humans, conceived primarily as individuals.

Mesmer's contribution was likewise to put Newtonian ideas to work at the human scale. Mesmerism was a holistic medical practice conceived in the last quarter of the eighteenth century, based on the concept of magnetic influence applied to human bodies and a theory of bodily homeostasis drawing on parallels with gravitational force and observation of the tides. It applied Newtonian ideas to human health and healing; the influence between planetary bodies could be replicated between human

---

[1] Peters points to Hesse (1961).

bodies, and physical relations repeated in social forms at the human scale. It gave rise to a modern theory of mind and of the unconscious and, over the century, helped shape psychological and social accounts of human relations.[2] As Peters comments, 'Mesmer and his disciples ... were offering a unified field theory of the material and moral forces. As gravitation held the planets in orbit, so animal magnetism kept souls in love and health and communication' (Peters 1999: 90). Yet, as a theory, Mesmerism contained the double possibility of mental harmony between subject and operator and the 'nightmare of loss of self to another's will' (Peters 1999: 89).

Animal magnetism brought its own vocabulary: the operator 'making passes', establishing 'contact' or a 'rapport' with the subject, which enabled 'communication': overcoming blockages in the patient's distribution of magnetism, forcing a 'crisis', and enabling the sharing of feelings and thoughts. For two souls *en rapport*, communication could take place over distance and, we might add, 'travelling' in the mind over both space and time, or clairvoyance, became a possibility.[3] Contemporary debates concerned whether the source of these powers was the manipulation of an (indiscernible or imponderable) physical property, or due to the operator's will, or, even, an expression of a common spiritual fellowship, a community of sensation that embraced all people equally (see Peters 1999: 92–93). It set an enduring framework, within which, Peters suggests, one can detect all 'the redemptive and diabolical features of mass psychology ... [including] the spell that dictators and admen cast on their audiences ... bypass(ing) ... the vigilant conscience of citizens' (Peters 1999: 93–94).

Technologies and spiritualism

This understanding developed hand in hand with the appearance of new technologies of communication throughout the nineteenth century, forms of reproduction and transmission which promised the meeting of

---

2   See Ellenberger (1970).
3   Balzac's *Ursule Mirouet* (1846) contains an early fictional account of such clairvoyance.

disembodied minds first separated by distance and then by time: the technologies of the telegraph, the telephone, the phonograph, daguerreotypes and photographs, film, and radio. 'Technologies such as the telegraph and radio refitted the old term "communication", once used for any kind of physical transfer or transmission, into a new kind of quasi-physical connection across the obstacles of time and space' (Peters 1999: 5). By means of electricity, communication could take place regardless of distance or embodiment, allowing the term to conjure up older stories, 'dreams about angelic messengers and communion between separated lovers', and this form of communication 'seemed far superior to the age-old ... face-to-face work of making lives together in language'. Communication in this sense appeared instantaneous, subtle, and capable of being experienced directly, without mediation. In retrospect, we might think the strangest thing about the concept of person-to-person contact of this kind was the complete effacement of the material apparatus required for mind to speak to mind in apparently bodiless communication; the technical infrastructure and the collective enterprise it demands vanishes without trace behind the appearance of minds touching. Yet this illusion allowed interpersonal relations to be 'redescribed in the technical terms of transmission at a distance – making contact, tuning in or out, being on the same wavelength, getting good or bad ... vibes ... Communication in this sense made problems of relationships into problems of proper tuning or noise reduction' (Peters 1999: 5).

Mesmerism travelled across the Atlantic and became established in the less formal New World society in the same period as the aftermath of the Second Great Awakening and the construction of the telegraph. There it gave rise to spiritualism,[4] 'the art of communication with the dead, [which] explicitly modelled itself on the telegraph's ability to receive remote messages' (Peters 1999: 94). The raps received were understood 'as a telegraphic cipher attempting to bridge the chasm between the living and the dead' (Peters 1999: 95).

The 'spiritual telegraph', with its mix of metaphysical and technical elements, was 'decisive in the making of the modern vocabulary and vision

---

4 See Podmore (1902) and Fuller (1982).

of communication', including developing the concept of the appropriate 'medium' through which to receive messages from remote sources. The techniques chosen to establish communication with the dead evolved rapidly, from Morse-like rapping to spelling out messages through a Ouija board, and then to automatic writing, direct voice, and a variety of physical manifestations and realizations of the dead.[5] 'Access to the hidden things of the spirit depended ... on having the right medium, and the revelations that came were grasped ... through the senses' (Peters 1999: 97). And new technologies were taken on board as they developed: spirit photography and recordings of spirit voices, not to mention automatic typewriting, to match automatic writing. In Peters' summary, 'spiritualism was one of the chief sites at which the cultural and metaphysical implications of new forms of communication were worked out ... and it is also the source for much of our vocabulary today (medium, channel, and communication) ... In spiritualism, "communication" was a concept that straddled the line between physical transmissions (the telegraph) and spiritual ones (messages from the other side). The spiritualist imagery of media is still with us today' (Peters 1999: 100).

Towards the end of the nineteenth century, a further mutation occurred and, in Peters' terms, communication 'acquire(d) its grandeur and pathos as a concept' (Peters 1999: 5). Earlier forms, the telegraph and telephone, had transmitted writing – Morse code – and the present voice of the speaker. Later technological forms separated from the limitations of speech and message, transmitting sound and visual images, with the potential for what we might call total recording, recalling stretches of real time in their fulness. That possibility represents the grandeur of these new forms. The pathos appears in the invention of two terms, 'solipsism' in 1874 and 'telepathy' in 1882: 'both reflect an individualist culture in which the walls surrounding the mind were a problem, whether blissfully thin (telepathy) or terrifyingly impermeable (solipsism)'. Communication in the modern sense has these two faces, 'the dream of instantaneous access and the nightmare of the labyrinth of solitude'. They set the frame for the contemporary period.

5   For examples, see Owen (2004).

In sum, each new form of transmission and recording ('the post office, telephone, camera, phonograph, and radio') promoted this concept of communication. At the same time, there is a second aspect to the history Peters sketches, and this is what he calls the 'spiritualist reception' of these new technologies. For each re-conception of the properties of matter disturbs the order of common thought and allows new improvisations; the technical exploitation of these new properties is received by the public mind and put to work, largely to moral ends, reconceiving human well-being and its ills and means of repair of those ills in accordance with the potentials revealed in each new account of reality. Moreover, these effects of reception feed back into and shape the way technologies are conceived and used; in practice, there is little or no space between claim and response. If the language of action at a distance is used to re-describe the variety of human relations in a new way, as if transparent transmission, mind-to-mind, were the norm and not the exceptional case in collective human endeavour, the language of those reconceived relations, drawn from Mesmerism, Spiritualism, Psychical Research and so forth, has equally shaped the vocabulary, concepts, discoveries and employments of the new technologies, circumscribing both their ambitions and their perceived failures. Metaphysical and technical terms are mixed together because the models drew upon a common source and then from one another.

The spiritualist inheritance

The third lesson Peters offers concerns the potential contained in this inheritance. To sum up the second argument, successive developments in the technology of recording and transmission lent support to the modern notion of communication, with its hope of direct contact between minds. Yet, at the same time, ignoring the complex material conditions needed for the functioning of the equipment meant this hope was experienced as much in its frustration as fulfilment. This two-sided experience of the frustration and promise of direct communication was given expression in concepts taken from the exploration of exceptional mental powers – of influence, 'travelling' and clairvoyance, and of contacting the spirits of the dead – which in their turn prolonged language drawn from early modern physics.

Every new technology of information storage and transfer then expresses this aspiration to direct – bodiless, mind-to-mind – communication, and has to be recovered and controlled by what we may call the human dimension: situated within an account that takes account of the reliability attributed to the speaker by the hearer, the feelings shared by the speaker with the hearer, and the world of understanding established common to both parties. For these conventions are disrupted with each technical innovation and the disruption expressed in, among other things, a desire for direct communication. In brief, technological changes concerning information storage and transfer will be accompanied both by the breakdown of established means of exchange and an increased desire for unspoilt communication. These are the experienced conditions of a shift in the conditions of knowing the world.

These alternatives are played out in each new technology of communication that emerges. Spiritualism played with the telegraph, Morse code and, later, the camera, but its ideas never became fixed; it continually transformed in conjunction with developments in contemporary materialist understanding. We can trace the changes in scale of the work of the spirits in Theosophy as they incorporated discoveries in the age, size, and atomic detail of the universe in the 1870s and 1880s. The formulation of 'phantasms of the living' and exploration of the nature of spirits by the Society for Psychical Research was a feature of the same period, contemporary with the new methods of recording and transmission in the phonograph and the telephone (Peters 1999: 141). Psychical research also responded to the radio (Peters 1999: 101ff.) and invented the notion of brain waves in the 1880s, taken over in turn by physiologists in the 1930s. And the appearance of spaceships, mainly in fiction but also in reports of sightings of 'airships', coincided with developments in film and radio in the 1890s, which also lent models assisting the evolution of psychical research into Parapsychology in the 1930s. Flying saucers, we might claim, are in this genealogical line that runs from Locke through Romantic recensions such as Mesmerism and Spiritualism, with a belief in direct communication together with a neglect of mediation (or rhetoric) and scale, focussing either at the individual scale or at the cosmic, but leaving aside the scale at which humans collectively make sense.

Peters sums up his notion of this tradition: 'In sum, a long string of notions, all of them dispensing with mediation and interpretation, invest the modern notion of communication: the spirit trumping the letter [*Pietist ideas*]; the immediate communication of angels [*Swedenborg*]; the "communication" of ideas via the medium of sensible signs [*seventeenth-century semiotics*]; the mind-binding power of animal magnetism; the lightning flashes of spiritual telegraphy; and the wireless rays of radio and brain wave ethers' (Peters 1999: 108).

And, on the other hand, the ideal of direct communication, of sharing understanding and instant sympathy, without reliance on words or speech, continually encounters obstacles – a break on the line, obscure messages, messages misdirected, overheard, or lost – and feeds off a longing to escape from a condition of plurality with fellow creatures whose perspectives are hidden from us and differ from our own.

Spaceships

From the beginning of the nineteenth century, then, each new medium that has emerged has been concerned with ephemeral communication at a distance on the one hand and archiving records on the other, activities which have created a 'parallel universe in which personal replicas dwelled and abided by laws other than those which apply to us mortals' (Peters 1999: 140). Each time, these replicas reorganize representations of the body and the mind and the possibilities of their interaction: the limits of distance and death (memory) are overcome, but at certain costs: a world of doubles without bodies appears, and a premium is placed on mutual presence 'in person'. These costs are played out in the paradox that new media, which claim to bring us closer, only make communication (in the sense conceived) all the more impossible. Hence the ambiguities of intentional creatures that do not speak, or do not speak to any effect – spirits, ghosts and poltergeists, aliens and flying saucers.

Peters divides his reading under the two headings 'recording' and 'transmission', although the topics are intertwined, as the first is mostly an account of the disappointment of expectations of presence in the obstacles experienced in the processes of recording, and the second concerned with hope of 'authentic communication' as the horizon of information transmission. If we for the most part leave aside his accounts relating innovations in media transmission and storage with regard to movements such as Spiritualism and Psychical Research, how does this approach concern flying saucers?

Novel forms of transmission (radio and radar) and transport (jets and missiles), combined with new forms of recording (film and photography) and manipulation of information (computers), together underwrote these appearances of intentional creatures who fail to communicate but who, nevertheless, monitor our activities, read our messages, and anticipate our thoughts. We might note, first, that flying saucers only appeared once radar had been developed and became confirmed when it was deployed in the States, with radar providing indirect 'evidence', and second, that the development of forms of transport – the actual transfer of bodies – played a role, 'travelling' being reproduced in other 'dimensions'. Alien forms were developed in tandem with the development of spaceflight, with the creation of the radio telescope, and with speculations concerning the nature of space and time, both outward (Cosmology) and inward (recovered memory). Aliens also prolonged occult forms (theosophical notions of races in particular – see Roth 2005) and adopted filmic modes of travel in time and space, with their abrupt changes and reversals.

Peters is concerned with the dilemmas created by new media, which, in essence, he sees as taking on two linked forms: are these reproductions democratic, spreading knowledge, insight, images, recordings across a wide public? Or does such dissemination represent a betrayal of private, personal (and possibly state) secrets? These issues are perhaps separable from breakdowns of the means of recording and storage, but Peters ties them together. Mostly, he says, media fail to deliver; there is not too much communication (contact between minds) but too little; contact is not made by the media and technologies fail to establish relations between bodies.

The appearances of flying saucers play out these failures in a variety of forms over time. First, contact is never achieved satisfactorily: the supposed event either does not take place – a missed rendezvous – or lacks evidence or record, given the failures and ambiguities of both radar and photographs (cf. the later repression of memory). Second, in the case of those claiming contact (known as contactees), the problems of communication may be overcome by telepathy, but this contact is at best banal and is unsupported testimony. Third, after the period of contactees comes that of alien abductions, where there is passing contact between bodies – probes, intercourse, implantation (all to do with DNA and the sequencing and remixing of information; see Doyle 2005) – but memory of these events is taken away and has to be reconstructed through hypnosis, and reconstruction through recovered memory places obstacles to any notion of direct experience or bodily contact. And last, we have the search for extra-terrestrial intelligence – the subject matter of this essay.

If failure is a recurrent theme of new media, this is because the system created (whether telegraph, telephone, letter, radio, or microwave signature) is a public one, allowing communication in general, but the messages conveyed are in principle person-to-person. The challenges introduced are all then to do with the breakdown of personal barriers – promiscuous connections, communications gone astray, privacy overheard or broadcast, potentially damaging materials passed under plain cover. Mediation – by media and signs – in short increases the potential for outside interference. Any act of recording, storage, and transmission of information (communication) allows the potential for the message to break free from its senders and receivers.

Having looked at the breakdowns in communication which result from recording, Peters turns to the search for 'authentic' communication in the promises of transmission: disappointment and hope respectively. Authenticity may be seen as the search for sincerity (cf. Latour 1993; Keane 2007) or for 'direct experience' in the Jamesian sense.

In dealing with the promises of transmission, Peters begins with the paradoxes of the notion of action at a distance (see Peters 1999: 177ff.). Action at a distance either is thought to work through some line of communication such as the ether – in which case there is effectively no action at a distance, only contiguity – or else all action is at a distance because there is always a space between bodies acting on one another. Communication is organized by these two possibilities (which we might call 'series' and 'structure'): either conceived as a stream of information unimpaired by distance or embodiment (as in Mesmerism, the electrical telegraph, Spiritualism, the wireless, telepathy, and so forth), or frustrated by the impossibility of communication occurring between bodies or minds (souls), which are always separated. Developing the latter possibility, there is a modern fear of isolation, a tendency towards solipsism: the notion that 'nothing exists save the projections of the self' (Peters 1999: 179), and that even touch is an illusion of the sense organs. Peters traces this 'cosmic loneliness' to Protestant moral responsibility and to Cartesian doubt. Solipsism is always the other face of the dream of communication without bodies. Both attitudes – the aspiration and the fear – bypass the body and are supported by modern conditions of communication. And the two mix: one may aspire to establish contact but is always beset by a sense of doubting its success. This duality appears in SETI and astrobiology alike.

Peters makes three general comments. First, that, if Eros is the attraction and repulsion between bodies that do not touch, modern communication is a 'record of the erotic complications of modern life' (Peters 1999: 180). He notes, second, how Anglo-American Idealism plays with these impasses, which are taken up by Eliot and parodied by Beckett. They are also challenged by Pragmatism: even if minds cannot know one another, solidary action can be attempted, taking time and establishing its own rules and activities (see Peters 1999: 183–184). And third, he moves to consider the need to interpret imposed by mass communications media: is this a message? Or is it inner projection? The interpreter has to bear the weight of the entire communications circuit and its failures, and any inability to decide is called paranoia: the psychological correlate of the social form of mass communication, seeking sense, determining what is a communication (a message meant for the listener, or intended to be concealed from him) and what is noise.[6] In sum, a modern Eros, a solipsistic way of life,

---

6   Compare the analyses of contemporary American culture offered by Lasch (1979), Rieff (1966) and Sennett (1977).

and paranoia are fruits of the concept of action at a distance and the concept of communication to which it has given rise.

II. Communication in the contemporary period

If we hold onto the hope of mind-to-mind communication, on the one hand, and the fear of messages going astray or simply silence, on the other, we have grasped the frame common to the two puzzles with which we began: the common pattern of seeking contact followed by diversion. But we have not finished with Peters' analysis, for he offers further insight into three relevant topics: first, the importance of radio to the imagery with which we are concerned, second, a series of observations on SETI within the perspective of the general condition of communication, and third, the existence of two periods in the twentieth century when focus on communication and its problems was particularly intense, which creates a time line allowing a review of exercises in Martian linguistics. His discussions give further substance to the concept of communication and point to themes that will be prominent in the writings to be considered in the rest of this essay.

Static on the radio

The radio signal was discovered in the 1890s; radio was at first considered uncanny, breaking apparently natural limits of space, time, and audibility, allowing voices to speak without bodies (Peters 1999: 211). Spaceships emerged in fiction exactly contemporary to the invention of radio and share in its uncanny nature. At the least, they express similar dislocations of distance, disembodiment, and dissemination. They are very like early radio: the ham radio receiver seeks communication from a distance and tunes in by chance; there are no obligations on either party, and particularly on the receiver, and there are gaps between transmission and reception; moreover, beyond human messages, there are other signals, interference or static on the radio (Peters 1999: 212). In similar fashion, the spaceships come from far away, travel fast, hide their traces and their content, and appear at any point. When they bear messages, the messages are general, concerning the perils of technology, criticisms of government, and desired reform of human ways. They spread their messages, but who will hear or respond? They promise bodiless communication – telepathy – but it always vanishes before it reaches us or reaches only the most unreliable witnesses. Even when they produce the appearance of one-to-one communication, choosing to whom they appear, the account is vague and the purpose of the visit unclear; these events may simply be static or interference, explicable by natural causes.

Radio was originally developed as person-to-person communication (Peters 1999: 206ff.). It took twenty-five years for the emphasis to switch from amateurs making point-to-point transmissions to the notion of public broadcasting, with populations in the 1920s listening to radio sets. This conception, of broadcasting speaking to a public square, was however always challenged by a series of moves designed to restore a semblance of personal relations – dialogue – to the medium; these included radio personalities, a conversational style, the production of 'soap operas' (serials with a domestic frame), the involvement of the public through live audiences, live broadcasts and the like, and the development of the listeners' participation through publications, news of the stars' personal lives, fan clubs, fan magazines and so forth. Moreover, the ideal of a public space constituting an informed, democratic polity was always undermined by possible threats: fraud in advertising, or the mimicking of sincerity by untrustworthy speakers, or manipulation of the listeners' sensibility and the potential for undermining authentic political or religious participation.

This frame played a part in the reception of flying saucers, which finally appeared in 1947. Their appearance was confirmed by the deployment of a radar defence system in the States, radar being in one sense a visual representation deriving from radio technology. The world they entered was that of the military-technological structure with its radar tracking stations, and saucers manifested both the disappointments and hopes of radio communication. The evidence they left was fitful, ambiguous, and unrecorded;

appeal was always made to the experience of the operators, who could be relied on, it was believed, to know whether the signal represented birds or a plane or an unknown. The debate was in terms of identification or unidentified, and there were no bodies or remains, only reports. On the other hand, the saucers also shared in the techniques of re-embodiment: they created dramas, direct communications and a sense of election; the 'fans' participated through various media (magazines, broadcasts) and testimonies, including flying saucer clubs, with 'contactees' as everyman or the public's representative; and the body re-emerged in the contingencies experienced, the sense of horror, threat and love evoked, the dramas lived through, and fears too of deception and fraud. Spirit-to-spirit contact transformed back into forms of touching the body, parallel to the techniques of radio broadcasting (cf. Peters 1999: 224–235) and, in each regard, flying saucers reproduced aspects of spiritualist practices.

The two forms – alien spaceships and radio – came together in the myth of the Orson Welles broadcast of *War of the Worlds* in 1938 (see Cantril 1940). Here, we find questions of direct voice – immediacy, live broadcast – and the contingency of tuning in – broadcast, but personal implication, overhearing a situation – and the effects on the boundaries of the person – threats of death, displays of fear and affection, the moving of imagined limits, the breaking in of fiction to produce effects in the everyday. In this transformation of the drama produced by broadcasters into unintended consequences in the listener, we have some clues as to the links between the forms of radio and alien invaders.

Peters' criticism of SETI

When Peters turns to the theme of 'communication with aliens' (Peters 1999: 246ff.), his argument is that SETI continues to express the dilemma of 'is it communication or is it static on the radio?' A good deal of his account is constructed around a comparison between SETI and the search for spirits.

SETI is 'perhaps the most sustained examination of communication – and communication breakdown – in late twentieth-century culture ... The project presupposes knowledge of the speed of light, the measurement of vast distance, the discovery of radio waves, means of sorting signal from noise (such as cryptography and information theory), high speed computers, and the longing to break through the circle of our own cognitions to touch otherness' (Peters 1999: 247). SETI is at the same time a philosophical experiment, an enquiry into our earthly dilemmas about communication, and an exploration of the consequences of the storage and transmission of information across vast expanses of time and space. And it is a doomed scientific enterprise, Peters claims, for it is a strategy seeking 'to transcend the inevitability of one-way communication' (Peters 1999: 246).

Why doomed? He offers two reasons. First, the obstacles of space: interstellar communication has to overcome a series of gaps: distances, delays, signal persistence through shifts, distortion, interference, and the prospect of radical otherness in such terms as intelligence, being in time, bodies, modes of communication, and so forth (see Peters 1999: 247). And second, problems of time: all such communications must come out of the past; all communication at a distance is communication with the dead and, simultaneously, we might add, with future forms of life, if we are dealing with civilizations more 'advanced' than the human.

Peters marks the parallels between SETI and spiritualism, as both represent reflection on the latest advances in communications technology, in each case engaging scientists at the margins of popular interest, and subject thereby to censure. He sees continuity between the two projects in their past Cambridge links, with Oliver Lodge and others working at the Cavendish Laboratory on radio emissions and radio astronomy. It may be possible, too, to connect with Myers' analogies of the subliminal consciousness with invisible parts of the spectrum, and also the project of 'cross correspondences', which must relate to radio waves. But, away from Cambridge, the notion of using 'radio as an instrument of communication rather than of inquiry ... appeared in the late 1950s, with Project Ozma' (Peters 1999: 249).

Peters identifies these features of Project Ozma which link the search for spirits and for alien intelligence: dealing with massive data, in the hope of making connection; driven by faith in the other's existence without being able to take hold of a sure connection; imagining the universe to be crossed by conversations we cannot tap into; needing technologies and methods to sort messages from static, such as Information Theory and cryptography; and also the need to control the tendency to impose patterns where there are none (see Peters 1999: 248f.). Peters mentions analogies with parapsychology in the same period, as well as the earlier parallels with spiritualism and psychical research.[7]

The limits of SETI are clear from the original article by Cocconi and Morrison. We assume that there is a community of intelligence out there, and that science is the universal language of that community. There is also the hope that we can enter into communication with that community – make our presence clear – and can receive help to overcome our present technological and ethical limits, avoiding the risks of destruction. To gloss his remarks, the suggestion then is that these older civilizations will have progressed far beyond our limitations, but in the same kind of direction, so that they can and will help us. Contact will then be visits by spirits from the future, a means of accelerating into the common Cosmic Mind.

The project therefore begins by listening for radio signals, seeking fragmentary messages. It then moves to identifying a potential listener, as it were at the other end, the emitter of the signals. The third move is then to send out messages, looking for confirmation or proof of intelligence. And the end stage is to seek junction or contact. This notion of contact relates to both time travel and to linking with the dead. We will consider some of the books which express these elements in the two chapters that follow.

This construction of listening and signalling has its own obstacles to overcome, principally in the search for signal rather than noise (the equivalent of sincerity), and in the precautionary fears of encountering unfriendly others (who might treat us, we might add, as we treat other forms of life – colonialism, exploitation of natural systems, farming, zoos, even extinction). But the project cannot be conceived of as dialogue, in Peters' account, even if conceived as signal/response, because the elements are separated: we have split apart attention (listening, even spying), hailing,

---

7   We might also notice the parallel debates in psychoanalysis about the place of telepathy, concerning whether transference involved elements of thought reading – see Devereux (1953).

recognition (imitation), and interaction. At best, galactic conversation can only be alternating broadcasts, exchanges centuries apart (he cites Lem's *HMV*). 'The fundamental problem of communication is not adjusting semantics so we mean the same thing with words', Peters says, 'but figuring out ways to come into fellowship with otherness' (Peters 1999: 252).

SETI expresses a constructed human desire to communicate with others, combined with an acute sense of the possibility of error. The sign of another mind is the mark of artifice, detection of intention, and the will to communicate. Peters points out this attitude reproduces the isolation and solipsism of the Idealist; the reason, however, why such a message has the potential to thrill is that it indicates a desiring body: 'Come here – I want you'.[8]

The alien mind has to be like ours to some extent – for example, using the natural wavelength of the hydrogen atom as the logical frequency to send an interstellar message, assumed by Cocconi and Morrison to be a universal constant (Peters 1999: 254). We are set the problem of how to recognize authentic empirical inputs within the all-colouring powers of human cognition. Because we assume nature to be without significance (noise), we are reduced to solipsism, for the only source of intelligible order is within us. Hence our sense of a communication breakdown: it is because we have reduced the realm of the sign to information.

In sum, SETI transposes a historical moment into travel across space, based always in a faith in science and 'high' culture. SETI is the essence of what 'communication' is: a denial of embodied-ness, of our being in mixes with other persons, creatures, and things. And yet 'communication' is also the means by which extra-terrestrial life inserted itself back into the life of the military-industrial constellation from which it had been expelled in the form of the interplanetary hypothesis.

Peters concludes with some reflections on the alternative to communication/information, or what is lost: we need to return to desiring bodies, or bodies whose presence we desire. We ignore bodies, and believe in minds

---

8   Hence, we might think, Nozick's 'RSVP' (Nozick 1985), and the 'Why the silence?' (Fermi) question (see Davies 2010). For a recent discussion of the Fermi question, see Conway Morris (2022), chapter 6.

which make contact, and exchange signs, and so we ignore much otherness within our orbit – life, intelligence, wisdom. SETI misses this 'sea of intelligence' (Peters 1999: 257): words and minds are part of the world, signs mixed with mortal life. This approach (that misses the point) particularly fits with an age when intelligence can be stored in media. This insight leads Peters to what we may call a rhetorical 'ethic': not fidelity to an original, but responsibility to the audience (Peters 1999: 266), not the touching of minds, but the coordination of behaviours.

The history of communication in the twentieth century

If the notion of communication emerged as a topic for reflection in the last quarter of the nineteenth century, Peters proposes that the concept took its present shape in two period of debate in the twentieth century, 'after World War I and after World War II' (Peters 1999: 10). In the first place, 'all the intellectual options in communications theory since that time were already visible in the 1920s'. He points to philosophical writings exploring the need to manage mass opinion, to social theories concerning crowds, the masses, and social classes, and to modernist literary works which portrayed the idea of a breakdown in communication. In the second place, such reflection appeared in quite different form in the late 1940s, taking information theory as the key. Debate in this later period looked in two directions, first, towards technical descriptions, reconceiving all sorts of disciplines in terms of information processing and, second, looking to therapeutic methods to resolve obstacles encountered in these reconceived areas of human endeavour, covering the spectrum running from personal relations to international dealings.

The 1920s

First, then, the 1920s. In this period, mass communication came into focus as posing problems for 'the future of democracy' (Peters 1999: 11), seen in such instances as the power of demagogues to exploit 'stereotypes, censorship, inattention, and libido', and in the ambiguous roles assumed by experts and political vanguards, necessary to mastering complex social processes but each with their potential to privilege their own interests or party organization, and to manipulate slogans and rallying cries.[9]

In part in answer to these threats, a second strand of intellectual debate saw 'communication as the means to purge semantic dissonance and thereby open up a path to more rational social relations' (Peters 1999: 12). Peters focusses on Ogden and Richards, *The Meaning of Meaning* (1923), who sought to separate the descriptive and emotive functions of language and to define a scientifically-based and unambiguous vocabulary – Basic English – which might free an educated public from the primitive 'word-magic' which believes that knowing the name gives power over the thing. By purging the active force of fictions from the public sphere, challenging rhetoric and persuasion, tradition and fashion, it might be possible to create a common consciousness which would allow a matching of minds. 'Whereas propaganda preyed on atavistic word madness, semantic analysis would provide a medium of communication for the needs of modern scientific men and women' (Peters 1999: 13).

Modernist literary works, in this analysis, exploited the other side of this worldview, seeing 'communication as an insurmountable barrier' (Peters 1999: 14); isolation and propaganda were two sides of the same coin.[10] He sums up the three themes as 'communication as the management of mass opinion; the elimination of social fog; vain sallies from the citadel of the self' (Peters 1999: 19). Each perspective combined the dream of perfect communication with the fear of solipsism and isolation, and even the dream could rapidly turn into a nightmare, with the possibility of mind control, on the one hand, and the irrational behaviour of crowds made up of isolated and alienated individuals, on the other. In short, a longing for shared

---

9   Serious engagement with these problems may be found in Weber's 'Science as a Vocation' (1918), Schmitt's *Political Theology* (1922), Freud's *Civilization and Its Discontents* (1929), and Husserl's 'Philosophy and the Crisis of European Man' (1936), all articulating a sense of international 'crisis'.
10 John Carey's *The Intellectuals and the Masses* (1992) covers this ground with a certain thoroughness.

interiority lent itself to a 'horror of inaccessibility', as well as to 'impatience with the humble means of language' (Peters 1999: 16).

Peters also looks for resources in the period which rejected the mentalist vision and the couple telepathy/solipsism, writers who looked to language and sociality as the core of being human. He cites Heidegger and Dewy as concerned respectively with 'the disclosure of otherness ... and the orchestration of action' (Peters 1999: 19), and proposes such concerns allow 'pity, generosity, love' (Peters 1999: 21) to re-enter the frame of social analysis. Within this perspective, correcting some of the excesses of communication conceived ideally as the meeting of minds, Peters mentions early media studies in the 1930s on broadcasting, advertising, and propaganda, and also the first engagements with popular literature as a topic in its own right.[11]

The 1940s

The discussion took on new form after the Second War when two styles of talking about 'communication' developed, a technical one about information theory and a therapeutic one about failures in communication and their remedy.

The first took inspiration from Shannon's *Mathematical Theory of Information* (1948), an account deriving from research on telephony in the 1920s and cryptography during the War. The theory cast information in terms of entropy, drawing the concept from thermodynamics, and gave a technical definition of 'redundancy'; it was concerned, in Peters' judgement, with 'signals' and not with 'significance'. However, as the model spread, 'information' became a substantive theory of communication of meaning as well as of channel capacity. Something familiar from the experience of war and of coping with bureaucracy and other aspects of everyday life in

---

11   He cites Gramsci and also Q. D. Leavis's ground-breaking study on *Fiction and the Reading Public* (1932); we might add Edmund Wilson's (largely unsympathetic) essays on popular literature in the 1940s (Wilson 1950) and Lovecraft's essay on 'supernatural horror' (Lovecraft 1945).

wartime – 'getting a message through', we might call it – was turned into a scientific and technological concept. And the understanding of information was changed: it was 'no longer raw data, military logistics, or phone numbers; it was the principle of the universe's intelligibility' (Peters 1999: 23).

In this expanded form, the concept became transferred to every scale of life, encompassing the idea of genetic information, the conception of neural systems as communications networks, the notion of enzymes and hormones as messengers, and the picture of the brain as processing information. Beyond biology, the unimpeded communication of information became a metaphor for sound functioning in marriage relationships, management, and international relations, and its failure an explanation for disruption and breakdown in those human relations. More, the 'production, manipulation, and interpretation of information' (Peters 1999: 24) came to organize a range of academic disciplines, from electrical engineering to cognitive science.[12] All human affairs could be thought of in terms of 'information, communication, and control', and information theory promised to unify all knowledge, just as, in the past, 'geometry, evolution, thermodynamics, statistics, and mathematical physics' had each promised.

Peters makes three general points about this understanding of communication as the exchange of information. First, this concept is close to Ogden and Richard's 'semantic' view and relates to the tradition of 'instantaneous contact between minds at a distance' (Peters 1999: 24). Put in different terms, information can be separated from all human contingencies: once distinguished from the surrounding noise, we can achieve minds sharing a single clear thought. This is the ideal of a scientific language.

Second, anything that processes information can be termed communication, breaking down a number of boundaries that had been used to distinguish and to make sense. For example, the notion allows the natural sciences to be put in a single frame with the arts – taking language as communication – and the social sciences – understanding communication as the basic social process. This is the basis of 'structuralism'. More, communication

---

12   Hayles (1999) tracks the history of Information Theory and its outworking in both cybernetics and science fiction, complementing Peter's work for the post-War period.

as an activity has ceased to be in any respect connected primarily to the human body, which in turn can become 'a site for exploring post-human couplings with aliens, animals, and machines' (Peters 1999: 25). In short, 'this view effaced the old barriers between human, machine, and animal' (Peters 1999: 24), central to the Cartesian account of human nature.

Peters speculates that this account in part derives its authority from its sharing a 'common cultural space and symbolism' with, not only the computer but also, in particular, the atom. Information can be spoken about in vocabulary borrowed from subatomic physics – its molecular or granular quality, its half-life and decay, its fission – and it shares the same 'semiotic spaces' – dealing in flashes, bursts, impulses, (mental) photons, the minimal quanta of mental stuff. And he concludes that 'both the bomb and information cater to a secret pleasure in possible apocalypse, the exhilaration moderns ... feel in contemplating self-destruction'.

Third, however, as if in compensation for this anti-humanism, there was in the same period a great expansion of the quest for authentic communication, for contact with other people. The post-War period was notable not only for the existential call for authentic disclosure but also for the cultural moment of communication as therapeutic self-expression. The therapeutic project is the second strand to the post-War focus on communication. It too reproduces the earlier semantic approach, seeing communication as 'a clarifying method that would work at both interpersonal and international levels' (Peters 1999: 26). Failure of communication was held to be the basic problem, which was then explored in the double optic of how damage arises and its potential repair. In the context of prosperity, commercialism, and television, we have repeated diagnoses of isolated selves and manipulated masses attributed to failures in communication or its wrong uses. Peters gives examples: Orwell's *1984* (1949), Reisman's *The Lonely Crowd* (1950), Wright Mills' *The Power Elite* (1956), Hoggart's *Uses of Literacy* (1957), Huxley's *Brave New World Revisited* (1958), and Williams' *Culture and Society* (1958); we could add the experiments of pulp science fiction in the same period to this list.

Peters attributes the roots of this vision of the ills of communication and their possible cure to the nineteenth century's disenchantment with Calvinism and its replacement with a therapeutic ethos of self-realization.

The vision was, however, given new expression by the invention of the technical strand of information theory; 'both the technical and the therapeutic visions claim that the obstacles and troubles in human contact can be solved, whether by new technologies or better techniques of relating, and hence are also latter-day heirs to the ... dream of mutual ensoulment' (Peters 1999: 29).

Against this view, Peters believes that 'the transmission of signals is an inadequate metaphor for the interpretation of signs' (Peters 1999: 30). In practice, human troubles of 'language, finitude, plurality' cannot be overcome by recourse to a bodiless conception of communication, and such recourse leads to longing for contact with spirits and other strange beings. 'The expansion of means', Peters concludes, 'leads to the expansion of minds' (Peters 1999: 29); new technologies are intimately associated with the generation of new forms of spirit life. He criticizes the therapists because they miss the facts, both that the self is centred outside itself and that signs have a public character. They imagine the self to be private, and subject to private experiences, and that language carries messages between these isolated selves; moreover, they reproduce this model in their treatments. And the technicians miss the limits or finitude of human endeavours, ignoring the fact that any repair to damaged communication will generate new failures, and that 'those who build new media to eliminate the spectral element between people only create more ample breeding ground for the ghosts' (Peters 1999: 30). In each instance, the proposed remedy creates more of the same problem; the antidote contains the poison.

To repeat a point made earlier, making the rare instances when we experience information exchange (or something close to it) the model and guarantee of the broader human goods of 'understanding, cooperation, community, love' is to reverse priorities and likelihood; it is to take the exceptional and derived as the key to the general state of things. Yet, when we turn to instances where languages have been imagined that allow commerce between humans and beings from other planets, they are all marked by this history. We shall reserve discussion of therapeutic languages; our present concern begins in the 1920s with the desire for a purified language that can by design change minds.

CHAPTER 3

Exercises in Martian linguistics

The concept of 'communication' in its modern form, particularly when condensed in the idea of 'information', gives shape to our understanding of how disturbances created by technical advances in the transmission and storage of signs are registered – monitored, made sense of, and put to work – and, at the same time, distorted, so that much of their significance may be missed. The two case studies concerning Project Blue Book and NASA are instances of this interplay of technical advances and the work of representation, engaging with both realistic and imaginative elements.

We now turn to specific instances of communication; fictional and non-fiction accounts of languages either coming from other planets or to be used in contacting other, extra-terrestrial civilizations. As we shall see, the literary examples focus around the ambiguities of communication, always stretched between the hope of contact between minds and the fear of silence, of hearing nothing. And although the designs for SETI languages are more pragmatic and less imaginative in their focus, for practical reasons, they too balance between the same poles. It is worth adding that, in addition to these two options, there are always hints present of an older, more modest account, of embodied participation as the ground of meeting rather than the aspiration to the mental exchange of information, but these hints are not our main concern.

As a general rule, these projects concerned with communication begin optimistically but turn in time to anticipations of failure. In this, they resemble the two case studies: hypotheses are tried out but then rejected, leaving traces that continue to play a role. We might note that Kittler (1987) and Peters (1999) offer two contrasting interpretative strategies in this regard. They share a common concern with the problems associated with information and disembodied communication, but Kittler focusses on the conditions of production of secrecy and on the potential abuse of information, while Peters is concerned with the *desire* for bodiless communication, in which minds meet unhindered by material limits. Both agree that the truth of this present period remains meaning embodied in particular persons and situations, shaped and mediated by language, and that the desire for bodiless communication (information) neglects the intermediate scale appropriate to human meaning making and ignores the specifics of each situation, imagining instead that meaning lodges simultaneously at the individual level and the universal, a product of the disembodied mind and, potentially, of the cosmic mind.

Kittler offers a pessimistic account of this situation, while Peters offers a more optimistic picture, not allowing the last word to the forms of misrecognition of our common situation which are generated from the development of certain features of that situation. He writes both of the aspiration to the ideal of direct (disembodied) communication promised by information technologies and of the experience of failure of these aspirations. Kittler is concerned exclusively with illusion, frustration, and paranoia, while Peters also notes the forms hope takes under these circumstances.

I. Technical innovation and science fiction

It is worth refining our ideas of what science fiction might be about in light of these remarks. Jameson (2005), following Suvin (1979), points out that the move to print was accompanied by the discovery of Utopias, and the introduction of new technologies of communication will be accompanied by new forms of Utopian speculation. Science fiction explores both dystopian fears and a strain of utopian optimism by considering multiple possible social futures. The combination of the frustrations and promises of disembodied communication set the forms available for much science fiction: these are the utopian forms in which the contradictions of the communicative situation and its limits are thought through.

In short, new technologies of communication produce new improvisations, for they promise bodiless communication (mind to mind), and yet such communication always has the potential to be overheard and dispersed and so to produce disruptive effects, affecting established powers, distributions, and hierarchies. Yet, simultaneously, the new technologies hold out the possibility of reintegrating those disrupted (potentially, all of us) into new political communities. There are linked movements of unbinding and of rebinding of imagined solidarities, which are connected to changes in the technical infrastructure, to human organizations, and to fictions.

This approach gives a sense of how in particular we might read science fiction dealing with life elsewhere, concerning both human life on other planets and the potential for an encounter with intelligent aliens.

We need first to be clear about the difference between speculation about and the practical reaching out to possible other civilizations. The practical history of human exploration of the possibility of life beyond the earth can be dated only from the end of the Second World War (1945). Although there had been interest expressed in such possibilities earlier than that date, exploration only became a realistic option with the extraordinary developments in a whole range of technologies during and under the impulsion of the War.

These technical developments are of particular concern in the present context because of one common feature: advances in technology undermine certain previously accepted forms of making sense in the world. Technical innovations not only make new things but also alter the conditions for making sense of novelties (and much else). In the previous, settled state, you could make sense simply by comparing the new-found to what was known; you could measure. In the new, disturbed state, you have to find the terms – the categories – by which to understand what has occurred. In shorthand, problems of measurement become problems of definition.

Science fiction may be considered as commentary on this feature of alteration in the conditions of making sense quite as much as it is an exploration of technical novelty – of things made – or of responses – continuities in human forms of life – to novelty. It is an epistemological genre as much as an ethical one.[1]

We have therefore two quite distinct social movements to consider: on the one hand, a technical complex, with scientists, industry and the military all involved, on the other hand, a science fiction milieu, with writers, publishers, activists and a wide audience of readers. The first is organized and fairly structured, not least because it demands considerable funding; the second is far more diffuse, though it too has certain organizational features. And the two are independent of one another, with their own concerns, rules, and times of development.

But they are not fully independent. In the first place, the science fiction milieu works directly with some of the products of the technical complex and puts them to work, in particular, means of information recording, storage and transmission such as film, television, radio, Information Technology and the Internet. The genre exploits the media through which the story is conveyed, and which allow its basic conditions of production.

Then, much of the basic subject matter of this imaginative work of writing concerns the hopes of communication created by these new technologies and the obstacles to communication experienced in their operation. In essence, the science fiction milieu explores the modern world – the 'contemporary', concerning what we are collectively on the edge of – conceived in terms of the exchange of information (including a picture of life conceived as the reproduction of information). For this kind of reason, the mind/body distinction and mental powers are a recurrent theme in all kinds of such writing. It would be possible to see science fiction as a meditation around the paradoxes of communication, with the elimination of the body and the freeing of the mind prominent themes.

And third, because of this central concern, the science fiction milieu in many respects offers the wider society, including members of the

---

[1] This draws a contrast with Jameson and Suvin, who both detect a direction to change and can therefore claim objective progress supports specific political positions. In practice, the disturbances associated with shifts in categories can result in many kinds of improvisation. See the discussion of anti-humanism in science fiction and its epistemological consequences, below.

technical complex, a chance to reflect upon itself: as well as including a lot of nonsense ('cod-science') and misunderstandings (which may themselves be enlightening), science fiction offers a kind of laboratory for scientific speculation and moral reflection, modelling the presuppositions at work in the technical sector. For this kind of reason, the science fiction milieu can play a role in such matters as shaping funding decisions and determining the plausibility of research programmes, as well as preparing the ground to accommodate new developments and even, at times, appearing to anticipate future innovations and events.[2] In short, science fiction exploits, explores, and helps to develop the technical resources of the contemporary world.

Various sub-themes can be identified in the materials to hand, themes such as time travel (or manipulations along the time axis), the work of deciphering anomalies and signs hidden beneath layers of concealment, a certain paranoia concerning the motives of large organizations, and utopian hopes of direct contact between minds, unimpeded by the constraints imposed by material conditions. All these topics appear to be effects of considering the modern world to be constructed above all around the communication of information. Yet, to echo Peters, while language is exclusively human, information is not; if we take language to be something like a part of how particular groups of humans achieve specific collective ends, information (in a Shannon-Weaver sense) is both more and less than that. Hence the particular interest of language as a topic in science fiction writing over the period; by focussing on language, and particularly the notion of an alien language, we can observe how authors and their audiences resist or accommodate the potential reduction of all human life to exchanges of information. I am interested in particular in instances where science fiction writers have conceived an alien language, looking at where they have drawn their models from and which characteristics they have chosen to emphasize.

It is worth remarking in a general way on the approach I take in reading these fictional texts; in a phrase, it is to read them with the same care and

---

2   A striking instance of the milieu making a direct contribution was the presence of science fiction writers on the 'Citizens Advisory Panel' supporting President Reagan's 'Strategic Defense Initiative' in 1983 (see Luckhurst 2005: 200).

attention as I apply to more technical, non-fiction writing. After an initial description, I identify which ideas catch the eye and demand further thought, exploring the connections they make to other texts and contemporary questions, and then ask where the models they deploy come from. The phrase 'catch the eye' points to the predominantly visual, filmic style of much science fiction, where everyday detail is the means of conveying meaning, rather than dialogue or plot. This process allows making wider comparisons, asking how the chosen texts contributed to the practical discussions taking place in the period, which are the subject of the essay. Starting from specific texts, we can both derive more general claims and offer evidence for them, in short, integrating text, history and ideas in a single argument.

Not all my materials are directly science fictional. I begin this survey by considering an attempt in the 1930s to propose a language adequate to the demands of the modern scientific world and trace its influence on one writer in the 'golden age' of science fiction. I then turn to possible successor models and their exemplification in two more recent novels, raising the question of which linguistic theories are being used. Third, I examine proposals put as part of the SETI project to create an artificial language which might allow communication with other minds or forms of intelligence, to see whether this kind of project is entirely distinct from or, on the contrary, overlaps with some of the fictional languages. And last, in a separate chapter, I review four novels devoted to 'first contact' that explore the constraints and possibilities contained in the SETI programme. My claim is that they are all framed by the desire for direct mental communication, on the one hand, and haunted by fears of the breakdown of such communication, or even silence, on the other.

II. General Semantics

The starting point for this part of the exploration is a book called *Science and Sanity*, published in the States in 1933 by Alfred Korzybski (1879–1950), who proposed a discipline he named 'General Semantics'. I rely on

Stuart Chase's popularization of Korzybski's ideas (Chase 1938), in part because he was a mediator of them for the science fiction authors I mention later.

Korzybski's thesis was that humans are affected by ills caused by their language; that, if it functioned properly, language would name objective reality; but that certain properties of language pervert the project, through use of convention, generalization, and – above all – abstraction. Abstractions tend to corrupt because their users suppose the often loose or even incoherent ideas represented by such words have a reality of their own.

This account bears parallels with other contemporary approaches, particularly Ogden and Richards (1923). It is in many respects a recapitulation of the ideas contained in *The Meaning of Meaning*, and 'General Semantics' bears comparison with 'Basic English'. Korzybski claimed that the structure of language carries traces of an earlier period, bearing vestiges of the anthropomorphisms and myths of a pre-scientific culture. As we presently have it, language is ill-adapted both for scientific research and for communicating the discoveries that are emerging about the nature of physical reality. And this poor adaptation has social consequences for, in particular, our language fails as a medium for education: modern children are not helped to develop in a way that leads to sanity, for their language fails to correspond to what we know to be the case.

This argument is then a strong version of the 'disease of language' thesis. The contemporary ills of society – delinquency, drug addiction, neuroses – and political ills – the rise of irrational ideologies of nationalism, Bolshevism and Fascism – and the ensuing conflicts and wars can all be attributed to deficiencies in our language. It is a strong version because it supposes human sanity is at stake: contemporary behaviour in the 1930s was marked by irrationality, the power of rhetoric over self-interest and the expression of political differences in terms of absolute oppositions, and so science became the key to a better world, not just in material terms but in those of reason and peace.

General Semantics invoked then both the extraordinary scientific and technical advances of the period and concerns about domestic and international unrest. It was thoroughly of its moment. The reason these external features were linked to the structure of language is simultaneously logical and physiological. On the one hand, Korzybski proposed that the human nervous system is naturally structured according to scientific principles; there is a congruence between the world as it is and its perception and representation. On the other hand, he suggested that nervous functioning may be damaged by poor (abstract) language which denies that congruence, leading to mental disorder. Language has power in the world, if only the power to cause damage.

The remedy then was to repair and develop language so that it corresponds to the physical structure of the world that was emerging in contemporary science; in this fashion, we may better understand both the world and ourselves. The world can no longer be thought to be ordered according to Euclidean Geometry and Newtonian Physics, and it can no longer be understood by traditional Aristotelean Logic: a logic of identity and non-contradiction is a good means of handling abstractions, but it is not a good way of dealing with reality as we now know it, where categories overlap. As Peters notes, various philosophers were concerned with similar problems in the same period.

In essence, Korzybski distrusted dealing with objects taken in isolation, and imagining that words name objects. He was particularly cautious with respect to the subject-predicate form: things have to be taken in context, in their contrasts and relations, instead of being attributed an identity of self. We should look for many-value as opposed to single-value appraisal, and an appreciation of relations, order, and structure (cf. *Gestalt*). In short, in Korzybski's account, the model for clear thought is mathematical rather than literary.[3]

Our greatest need then, in order to achieve social reform and harmony and to educate ourselves and our children, is to create a language which resembles mathematics, and so relates harmoniously to the structure both of the nervous system and of the environment. Yet most modern languages – with their use of the verb 'to be' (promising identity), their taking nouns to name relations and actions, their capacity to confuse levels

---

3   The proposal that we should discard historical and anthropological methods in favour of a cognitive science approach is a contemporary version of this model; see the discussion in Section V of this chapter.

of abstraction and to offer single value judgements – are unlike either our nervous systems or our environments.

Korzybski founded the Institute of General Semantics in 1938 to promote a series of educational reforms to help meet the present crisis: first, a better general understanding of mathematics, to allow the grasping of relations and situations; then, a constant trained effort on the part of individuals to cease to identify things with their descriptors and, in particular, to be sparing and self-aware in their use of abstractions; and last, a seeking to reshape language habits, identifying levels of abstraction and contexts of use.

Without these precautions, he believed, most discussions end in emotions, not rational argument. With them, there can be therapeutic, educational, scientific, and political clarity. Communication can be established, and communication failure (a confusion of inference and description) avoided. There is no moral principle at work, but simply an objective discipline of finding the object or referent to which the thought and word refer and establishing its attributes and relationships.

I shall avoid making any criticism. The point at issue is that this theory was taken up with enthusiasm by a series of science fiction writers in the period, including A. Van Vogt, Robert Heinlein, Frank Herbert, and L. Ron Hubbard.[4] With respect to Hubbard, we might note in passing a possible source of auditing, engrams, and clearing. In Herbert, we find the clue to the powers of the Bene Gesserit way. And Van Vogt's work deserves re-reading in this perspective, starting with his method of automatic writing and focussing on the non-Aristotelean logic ('Null-A') to which he seeks to give expression. But I shall concentrate on Heinlein, who picked up Korzybski's educational programme in detail and sold it to the Martians.

Heinlein's Martian pedagogy

Heinlein takes the notion of a well-ordered language which can remedy contemporary human ills and constructs an account of the conditions

---

4 Luckhurst offers valuable commentary on Korzybski's influence on these authors (Luckhurst 2005: 74, 93, 161).

needed for this language to be taught successfully. This model appears in his 1960 novel, *Stranger in a Strange Land*, taking the form of the Martian language from which the hero, Mike, gains his strange powers. The book has sold over five million copies, and there are groups which attempt to live out its ideas (see Cusack 2010).[5]

Heinlein had taken a course in General Semantics (and read Chase), and the novel can be read as an elaboration of Korzybski's concerns, both in the broad lines of the account offered of American society and in the remedy proposed of an integrative, logical language. Mike's Martian tongue represents an idiosyncratic version of a perfected General Semantics.

Mike is a human brought up on Mars and reimported as a young adult to Earth. The human world he discovers is controlled by two preoccupations, sexual desire and fear of death, and these are mediated by religious communities. Martian culture is differently ordered: religion, science and philosophy cannot be distinguished, and young Martians are formed by work done to perfect the self through acts of will while, at the same time, seeking to construct ethical relations with others. Martians pass through a series of distinct stages in their life cycle – egg, nymph, metamorphosis into the adult, and translation to ancestor – with the focus being on the life of the mind rather than reproduction in the adult stage, and a continuation of this mental life beyond (voluntary) 'discorporation'. Because of their life cycle, Martians are not driven by sexual desire to the exclusion of all other ends, nor do they fear death.

As an aside, this form of life bears a strong resemblance to the ambitions of liberal Protestantism, fusing an understanding of science, religion, and human potential in a form of this-worldly salvation. The realization of these ambitions in the Church of All Worlds, which Mike founds, belongs to what Albanese (2007) calls American Metaphysical Religion, and Kripal (2008) 'the religion of no religion'. Heinlein's account repeatedly evokes tropes of New Thought, Christian Science and Scientology (and many others, such as the Oxford Movement).

---

5   On Heinlein, see Franklin (1980); on *Stranger*, see Roberts (2016: 337–338), who cites a number of critical appraisals.

In the Martian life cycle, much depends on the qualities of mind of the discorporate Old Ones, effectively spirits who communicate by telepathy. Because of this mental property, deception is impossible on Mars, as the highest minds are permanently open to one another; indeed, they collectively form a parliament of minds or a super-mind. Less mature Martian minds – those of earlier stages – simply receive their appropriate instruction, taking the teachings and decisions of the elders on trust. The mental powers of even immature Martians, however, far exceed human capabilities.

The gift Martians can bring to humans is to teach this mutual transparency of minds, for with it comes a combination of innocence, insight and honesty, an integration of advanced science with wisdom (as opposed to the present human combination of reductive materialism with hypocrisy). Yet Mike finds his gifts an obstacle at first, for humans do not immediately intuit one another (a holistic understanding which is termed 'grokking' in Martian) and are capable both of deception and cruelty. Moreover, there are complexities, for Mike discovers that, because of their mental privacy and the dilemmas this restriction confronts them with, humans have developed notions of freedom and choice that are absent on Mars – but we will leave those issues to one side.

With focus on the mind rather than the body, a Martian upbringing has given Mike all kinds of paranormal powers: control of bodily metabolism; the ability to depart from the body and bi-locate, and to alter the dimensions of the body; the power to cause mechanisms to fail and things to move or even disappear, and the same power over people. Once Mike has learnt to understand human limitations and insights, he can control minds, read their potential, and foresee events.

Moreover, he can teach these powers to others through the medium of the Martian language, which is a great deal richer in the appropriate terms and ideas than English. He sets up the Church of All Worlds in order to recruit and form a new human race and, as one observer states, 'The Church of All Worlds is not really a church … primarily it's a language school … to teach the Martian language'.

Progress in the Church is controlled by competence in Martian, and there is a series of stages, like degrees in a lodge; it has the structure of a secret society (cf. Jenkins 2013), although the structure (and promotion) is guaranteed by the telepathic insight Mike possesses into the minds of the neophytes, calibrating their development. The end result is a group of people freed from the here-and-now, seeking truth as matter of fact, and putting truth to practical purposes, making war, hunger, violence and hate unnecessary. The only hitch is to find recruits who are willing to put in the work necessary to learn the Martian language.

But what are the characteristics of the Martian language? Following the themes of General Semantics, its structure will correspond simultaneously to the structures of the human nervous system and to the structures of the world. It not only serves to describe, but also helps to form, the world, through the harmonious participation of the actors; physicality, insight and imagination all interact. In Martian, one can aspire to the Word forming the World.

We are not given any linguistic description, but we are given material on the central term 'grokking'. We are introduced to its multiple senses serially – as the penny dropping, then intuitions as to the intentions of others, then comprehension of conversational and social context. We then move from 'normal' mental powers of receptivity – of intuition, understanding and sympathy – to 'supernormal' mental powers of telepathy and telekinesis. Because of the possibility of the intercommunion of minds these properties allow, the term extends to the collective deliberations of the Martian Old Ones, and also their powers of mind over matter, including the ability voluntarily to 'discorporate' or die.

Applied to the human race, Mike can see that the essence of humankind lies in the power of sympathy extended in this fashion; through it, humans may participate in what may be called a Cosmic Soul or Oversoul,[6] realizing themselves as part of all animate life, extending the perfect sympathy that exists between Martian brothers beyond any limit. They also may gain the mental powers associated with this extended condition. Hence the greeting used between initiates, 'Thou art God'.

I have underplayed the sexual themes present in the book, but you can see how ideas such as the overcoming of appearances and conventions, the fusion of body and mind, and the mutual participation of minds, can readily

---

6 Though I do not think Heinlein uses these terms.

find their expression in sexual relations; Heinlein is much of his period in exploring these notions. Sexual jealousy comes to stand for all the wrongs humans inflict on themselves and each other through their habitual lack of sympathy. In this characteristic, too, Heinlein's description recalls characteristics of many radical religious groups in the Modern period who have sought to live out the life of the Kingdom of God in the present time. If the failing particular to jealousy is the desire to close down communication and to keep relationships exclusive, with the mutual transparency of minds, lived in the present moment, both deception and restriction become impossible. Instead, there is a regime of both total honesty and promiscuity, which offers an interesting side light on a transitional era coming to be dominated by the concept of information.

We should also note that, beyond this temporary engagement with bodies and their mutuality, the horizon is one of discorporation and full participation in bodiless mental engagement, without impediment or delay. Mortality, sexuality, and time are swallowed up in a single moment outside time, when we shall know each other fully, mind to mind. This sounds uncannily like a Gnostic vision of a Christian heaven, where bodies are set aside.

Heinlein's novel belongs in a sense to the satirical genre of *Gulliver's Travels*: it draws attention to aspects of the human social order that stand in need of reform, and offers a response – the Martian language, in this case – which is not meant, perhaps, to be taken seriously. Or at least, its serious intent is concealed behind irony. The author draws on Korzybski's analysis and proposals for reform which, however, do not hide behind such a surface.[7] The one thing we can draw from this construction is that Martian forms respond to human concerns: we never leave this world.

In sum, this first attempt at construing a Martian language is constructed around the desire for transparent, mind-to-mind, communication, and the frustrations and impossibility of such an aspiration are paid for by the human side, which is nevertheless supposed capable, under instruction from another

---

7   Jean-Luc Godard's film *Alphaville* (1965) contains a critique of a science-based attempt to reform the world in the direction of utility and reason by control of its vocabulary; the headquarters of the dictator of Alphaville (whose name is Von Braun) lies in the Institute of General Semantics.

civilization, of transcending its limits. The group of writers mentioned earlier are distinguished by the clarity with which they anticipate the possibility of exchanging information, together with acknowledging the reforms that would be needed to put ordinary language to the service of this project.

III. Later developments

Heinlein's story is firmly anchored in the debates concerning direct mind-to-mind communication that emerged in the 1920s and 1930s. While not unconcerned with bodies, its focus is on the transcendent properties of mind, with contact with the dead, powers of telepathy and clairvoyance, and the ability of thought to effect physical changes all prominent. Korzybski can be claimed as a key player in the golden age of science fiction, in large part because he offers a model of a language constructed along logical lines (and so believed to conform to nature) which succeeds in integrating the mental and physical aspects of life. The potential of this language emerges first in mental gifts, but clearly also may recast reality in a non-trivial way. The model proposed, however, is very crude from a linguistic perspective, and General Semantics is not thought by linguists to have much to do with their discipline of 'semantics'. I want now to turn to two more recent science fiction texts which offer a more self-aware reflection on the models and powers of language which might be at work, and which are products of the later – post-1948 – information-focussed period of concern with communication, before turning (in the following sections) to the unique problems that confront languages designed potentially to realize contact in reality with another planetary civilization.

Is Chomsky or Whorf a better clue to alien languages?

The fictional texts are Samuel Delaney's *Babel-17* (1967) and Neal Stephenson's *Snow Crash* (1992). They have been paired by a linguistic anthropologist, David Samuels, because they exploit respectively particularist and universalist models of language – Whorfian and Chomskyan accounts (see Samuels 2005: 124). As Samuels writes, the issue in imagining alien tongues lies in the question of how languages and realities are connected. We may be able to communicate if the alien and human languages share a common (deep) structure, or we may have a chance of doing so if we share the possibility of similar life experiences (Samuels 2005: 122). Is the basis of communication to be found in the internal structure of the languages, in which case, we shall probably be looking for how information is conveyed and how obstacles to communication may be overcome? Or is it to be found in the relationship between 'languages and the minds, cultures, and experiences of their speakers' (Samuels 2005: 124), in which case, we may be more concerned with local projects involving cooperation? This question echoes aspects of Peters' distinction between two models of communication, one focussed on information, the other on language.

The subject of Stephenson's *Snow Crash* is not an alien language precisely but an artificial one: a virus which not only causes computers to crash but which also crosses the screen, leaving the virtual world to affect the nervous system of the computer programmer. Stephenson thus succeeds in connecting the virtual and the real (which is something William Gibson, who invented the notion of cyberspace, notably failed to do in his earlier novels). The symptom of those infected sounds like glossolalia, which comes from 'structures buried deep within the brain, common to all people' (Stephenson 1992: 192). The backstory is that the diversity of languages was itself the product of a virus, and this return to an earlier hard-wired unity is expressed in metaphors of code-switching: swapping between languages in the context of globalization and the adoption of English as a global language (see Samuels 2005: 125).

Samuels, however, leaves to one side the role Stephenson attributes to neurolinguistic programming, a 'science' developed by Bandler and Grinder in the 1970s which bears certain resemblances to General Semantics. The reason hackers and programmers are particularly vulnerable to the virus is that through code learning they have formed neural pathways in the brain through which the virus can convey itself in a holistic down-loading comparable to intuition. The human brain (taken to be equivalent to the mind) is itself an operating system. Chasing down the functioning of the virus involves neurolinguistics studies and tracing its earlier history (in which the Tower of Babel is emblematic). This is an account of a hypothetical universal original language, the language of creation 'when naming a thing was the same as creating it' (Stephenson 1992: 260), a power which is brought into relation with computer languages. The virus allows an operator to take over the minds of others, just as hacking allows the control of computers: both taking control at a deeper (and, in the human case, a more primitive) level. In short, 'with a little venture capital, this neurolinguistic hacking could be developed as a new technology which would enable [one of the story's characters, X,] to maintain possession of information that had passed into the brains of his programmers' (Stephenson 1992: 377).

Information technology, in this fictional account, explains the power that one mind – that of the entrepreneur – may exert over others – the programmers; it exploits one of the fears raised by mind-to-mind communication, the taking over and capture of minds. It is based on a myth – the power of the Word – which both explains the origin of human woes – the Fall – and allows for the possibility of human redemption.

To achieve this account, we are however offered a series of reductions, of mind to brain, of brain to computer, of language to information or programme, and of the diversity of languages to a single origin, the origin of what is conceived of as the human biological information processing system. Whether or not it is fair to attribute this kind of approach to Chomsky remains an open question.

As with Heinlein, where Martians remain off-stage, Stephenson's sophisticated ideas about languages are closely earthbound; there is not an alien in sight. Aliens indeed, or alien machine life, generally only need appear in order to allow human problems to be thought through. Stephenson's concern is with the developing world of the Internet, together with a futuristic account of dystopian megacities. Aliens however reappear (off-stage again) in Delaney's novel, which plays on several of the same themes (and which bears the marks of *Stranger* in some regards).

In Delaney's account, 'language is also a virus – one that changes the thought processes of the person immersed in the language … The protagonist … is asked to decode some [intercepted] enemy transmissions

... As it turns out, the transmissions themselves are the weapon. As you learn to think in Babel-17, you are compelled to behave in certain ways' (Samuels 2005: 125).

We might notice that this feature of language resembles many cultural practices: religious, political, and economic ideas, for example, may involve aspects of their own realization, hence their contagious (virus-like) quality. In many things, we do not first consider the options and then choose whether to participate or not but are caught up in shifts that alter perception and action; recession and fear of foreigners can possess this social aspect, but so do appearances of spirits, aliens, and mental influence.

Samuels sees Delaney's approach as a 'strong' version of the Sapir-Whorf hypothesis: 'the idea that languages determine the thought of their speakers'. Reality, we might say, in this account is 'modal' rather than 'indicative': indigenous classifications define, order, and evaluate 'things' as desirable, optional, necessary and so forth, or their opposites. Here the real distinction (to develop Samuels' argument) is between what is permissive (matters of possibility and choice) and what prescriptive (matters of necessity, which cannot be negotiated). In Samuels' version, 'we experience and operate in the world because of what our languages make possible'.

The protagonist discovers that Babel-17 has no first-person singular, and so no individual ego to assert any thought outside the given linguistic framework. It induces loyalty to the alternative social project: 'It "programs" a self-contained schizoid personality into the mind of whoever learns it, reinforced by self-hypnosis – which seems the sensible thing to do since everything else in the language is "right", whereas any other tongue seems so clumsy' (Delany 1967: 189).

In Delaney as in Heinlein, introducing a new language comes down to achieving control over other minds, 'educating' behaviour through adjusted concepts. In an interesting passage concerned with how to resist the influence of Babel-17, the issue remains one of who has control over other minds but shifts in part from language to incorporate mental influence. We learn that 'the human nervous system puts out radio noise' which is received by other humans, so that someone with exceptional control over their thoughts and language – a poet in this instance, rather than a Martian – can affect those with less control. We also learn, however, that the reason this competition between languages is possible is that the aliens were originally an offshoot of human civilization, so we return to inter-comprehension or even competition being possible because of a single origin. This account then skirts around the important question of whether intercommunication would be possible between truly distinct species, with no common origin.

The opposition between a deep structure or, on the contrary, cultural categories as the source of language's powers of communication and intelligibility, though important, then seems less significant in these stories than an underlying notion of a 'true' or scientific account, matching language to reality, and capable of adjusting reality through a perfected language. Language is seen by each author as speaking directly to the mind and as being open to being put to work by an exceptional person either for truth (science) or for ideology. In this regard, it is worth noting that Stephenson alone sees a danger in these theories and ends up with a defence of a plurality of languages as the only guarantee of human freedom. Neither author, however, glimpses the rather different thesis concerning language that Peters puts forward: that language is one of a number of means by which humans may undertake cooperative projects together, and indeed that it may not feature much in collaboration between species.

Both writers, then, share aspects of Heinlein's 'Gnostic' view of language. Such a description does not matter with respect to science fiction novels, which are entitled to their unorthodox speculations; that indeed is their justification. But the question of 'orthodoxy' does matter when confronted with real world problems, for an 'orthodox' account is supposed to work better practically (in the long term) than an unorthodox thought experiment. It matters getting your categories right. So, the kind of language model adopted has implications potentially for real world projects hoping to receive communications from other planets and perhaps to send them in return. These projects too have developed 'languages' which it is hoped may allow us to send messages to distant listeners, without distortion or particularity, and likewise have developed procedures to decipher any communication they might have for us.

Anti-humanism as critique

Before turning to these practical languages designed for interplanetary communication, it is worth taking further the core issue in the novels discussed, which is both the distance we may anticipate between human capacities and those of the alien, and the effects registered by each party, particularly on the human side. In these three stories, alterity – the alien – takes different forms. The aliens are effectively spirits who offer telepathic powers in Heinlein, hidden alien messages with the capacity to alter human minds in Delaney, and human coding that can control minds and behaviour in Stephenson: spirits, aliens, and artificial intelligence. But in each case, their effects are alike: contact with the other produces new mental capacities and hence new behaviour within human populations.

Alterity as such is the ostensible subject matter and may find a wider range of representations; by reading other stories, we could add further candidates: animal intelligence, enhanced humans (cyborgs), and artificial mechanical forms (robots). Science fiction has explored all these 'icons' (Vint 2014) of alterity, which have escaped from their original contexts in stories of one kind and another and have become part of widely used resources for thinking about the present world. But the deeper concern in each case is the alteration contact with alterity produces within humans: we undergo a change in our human nature into something rich and strange.

In practice, stories about alterity cover a range of changes. We used to consider animals and machines as other than human; this is at base the Cartesian move defining the boundaries of the human. But now these boundaries have become blurred; we want to know how much of the human is shared with animals and machines (computers in particular) and, by extension, what to make of alternative forms such as aliens, angels, and spirits. More confusing still, humans can be altered in reality by genetic engineering, biological and mechanical enhancement, and crossing the human-computer interface, hence the interest in transhumanism. These possibilities raise the question about the presence of the 'inhuman' within the human (Lyotard 1991) and the potential for the partial or wholesale replacement of humans by biological hybrids or engineered machines, from which talk of the 'posthuman' develops, covering both the spectrum of possible outcomes (see Hayles 1999), and the 'Singularity' when this replacement shall occur. The problems of communication and their effects on the recipient lie at the heart of all these discussions, and science fiction is a key resource for such thinking because we live in a world which has become 'saturated' with technologies (Luckhurst 2005).[8]

Elana Gomel's *Science Fiction, Alien Encounters, and the Ethics of Posthumanism* (2014) offers a good outline of the issues. Its particular strength is that it identifies the wider context of these debates, which is the longstanding discussion about the adequacy of humanism as a philosophy in the context of the twentieth century. And its interest is the thoroughness with which she pursues the option of anti-humanism as a key both to the potential of science fiction and for understanding the contemporary world.

It must be said at once that very little science fiction fulfils the brief she holds out for it, other than in hints. And this is true for the texts I consider, both fiction and non-fiction, for the ambition of communication constantly raises the question of whether sender and recipient of a message might share some common understanding. Nevertheless, her discussion is of value, highlighting the limits of humanism and asking whether science fiction in general desires to go beyond humanism, despite the array of icons of alterity it mobilizes, thereby offering new clues of what to look for in these texts.

Her consistent target is humanism, and her book an exploration of its end and seeking out its avatars. She cites Vint (2007), that technology makes the concept of the 'natural' human obsolete. Hence the recurring theme of the transhuman or posthuman. But Gomel's narrower target is the persistence of humanism in the text, marked by the Golden Rule, at once

---

8   Or, at least, we have a literature which takes this change for granted. The importance of an academic discipline studying science fiction cannot be underestimated. Since the 1960s, there has been a sufficient concentration of scholarship to offer a new context to which the writing of science fiction responds, adjusting and multiplying its categories, and which has spread the influence of the genre in society, affecting media and wider concerns. This is not the same thing as the 'saturation' of the modern world with technologies; rather, there is a feedback loop which plays shifts in literary understanding and ideas as mapping real-life conditions, instead of adding glosses to longer-term continuities.

ethical – 'Do as you would be done by' – and epistemological – focussing of sympathy and identity – so that ridding us of the Golden Rule demands new forms of subjectivity and narrativity. Her hope is that different stories of alien encounters will embody different kinds of response to the presence of 'radical alterity in the Universe' (Gomel 2014: 4). And her fear, that little in fact changes, and that all we do is meet our unregenerate selves in these new forms.

This project is then far-reaching, for it demands not only decentring the human in thought but also criticizing the real-world humanistic framing of 'family, ethical and political behaviour'. The other must be understood as a moral agent 'but with a morality different from mine' (Gomel 2014: 5). In short, aliens are not human, so what are the appropriate 'inhuman' responses, once we have had to discard 'tolerance' and 'respect', and even good and evil, in an effort to think their difference? And this thinking has a knock-on effect, translating into considering what forms of agency are possible with our posthuman bodies and, not simply new bodies and selves, but also what cultural and civilizational changes come with this 'otherwise'?

Her thesis, in sum, is that, with technology, we find humanism does not work and we use aliens to imagine the rules of a posthuman existence (beyond good and evil). Citing an essay by Lem (1984), 'we need aliens because we are already alien to ourselves' (Gomel 2014: 4). And her prescription, that we should abandon the Golden Rule and discard sympathy and identity, and, in their place, look to difference, incomprehension, and alterity.

Gomel's research question is 'What work does 'the alien' do in science fiction literature' (confining herself mostly to the printed word)? The area she looks to is that of literary technique, 'narrative tools, conventions, and meanings' (Gomel 2014: 3). And her answer, also drawn from Lem (1984), is that we engage with certain theological resources, while stripping them of religious ideas and implications.

She starts from the assumption that we (humans) have a profound need to know we are not alone in the Universe (Gomel 2014: 8–9). Then, she asks whether we are looking for reflections of ourselves or, on the contrary, for notions (rather than evidence) of different kinds of intelligence, in order to challenge our assumptions about ourselves. This distinction allows her to classify science fiction works broadly as seeking sameness, whether in conflict or reconciliation, on the one hand, or transcendence, on the other.

The key failing of the first option, whether the aliens seek conquest, companionship or merging with us, is anthropomorphism – a variant on the ethical play of similarity and identity: they are like us. The analogy is excessive because of the restrictive model of humanity that is applied; the ambition is to shore up the boundaries between the human and the animal, on one side, and between the human and the machine, on the other, a belated defence of Cartesian difference.

Beyond the couple of mutuality and conflict, however, lies the category of 'transcendence', which is why Gomel's approach can draw on religious materials. She resolves the trope of the opposition between the sciences and religious forms of life by the appeal of both to wonder, awe, the sublime. At the same time, she distinguishes between naturalism and supernaturalism: aliens are natural beings, neither angels nor demons (Gomel 2014: 13). Then, the challenge of representing the truly alien mind is comparable to that of representing the divine. She cites Lem again: science fiction can avoid anthropomorphism by borrowing tools from Theology and Mysticism. Look to the poetics of the numinous and the sublime, Lem says; aliens are not a metaphor for God, rather, God becomes a metaphor for aliens (Gomel 2014: 15). We should borrow from what is called 'negative theology'.

A word of caution: Gomel is pursuing some interesting lines of enquiry, but, first, the assumption of a natural religious instinct is without evidence (she wrestles later with the universal claims of cognitive science and whether it may support a notion of the 'sacred') and, second, the notion she imports of negative theology is suspect. In her interpretation, negative theology supposes mystical 'experience', experience which is specialized and rare, and it is concerned with this rare encounter with the non-human. In the tradition, however, negative theology is about ordinary language and concerns how we can talk about the encounter with a 'difference that is beyond difference', rather than how we might assimilate that difference to some class of our experience. Gomel is on to an interesting problem; the important point – or the starting point for a critique – is that negative theology is about the insufficiencies of language for grasping things found in everyday life, not about exceptional experience. And we might suspect that the appeal to 'the numinous and the sublime' opens the way to humanism's return.

Nevertheless, the insight is important: negative theology is a key theological tool we can borrow along this line of argument, suited to the postmodern period when we look particularly to difference, absence, and otherness. And she identifies this shift with talk of the sublimity of the cosmic order, revealed by the sciences. 'Stapledon [she cites by way of example] reaches for a kind of poetics, in which disintegration of language, obscurity, and silence are signposts towards the sublime' (Gomel 2014: 18). Gomel's positive contribution is founded on this play between two kinds of religious approach, positive belief on the one hand, which she rejects, and negative theology on the other, the latter the apprehension of a completely other or alien.

If that is Gomel's theoretical contribution, what consequences follow from this initiative? She identifies three false directions, or paths she intends to avoid, before turning to some positive proposals.

First and most important, she rejects alliances with other critiques of humanism – socialist, feminist, animal rights – which, in each case, are caught in the dilemma posed by adopting the humanist value of emancipation. Time and again, criticisms of the present order, whether of colonialism, class struggles, gender inequalities, racism, sexual prejudice, or the exploitation of animals and of nature, seek to reimport values of fairness and justice which repeat the original prescriptions of the dominant society as the agent of civilization and progress. All our spontaneous ethical impulses are rejected.

Second, she asks whether we can rescue ethical thought by recourse to an updated Darwinism, looking at some recent attempts to create a 'science of good and evil'. Her basic criticism here is that the naturalism of 'Theory of Mind' restates the Golden Rule, for likeness and empathy are taken to be fundamental ('hard wired'). It therefore fails to understand the other. We extend likeness to disenfranchised groups, to embryos, or to the Biosphere (Gaia), and then debate which aspects are sufficiently 'like' to benefit from our compassion, identity and so forth. Thinking about altruism and sociobiology fall under the same criticism. Gomel concludes that we cannot defend likeness sufficiently to justify an evolutionary basis to morality.

And third, when she considers attempts to found ethics in alterity, she concludes that, although indicative, they all ask the same humanist questions.

The positive face of anti-humanism

Unlike animals and the Biosphere, aliens represent entities that are moral agents in their own right and yet do not know the Golden Rule. They therefore allow us to confront the limits of humanism and our current political and ethical crisis. Gomel asks, what is our human response to them, beyond empathy and a sense of unity? She looks to new forms of narration and representation that create a sense of 'distance' and 'alienation' and 'fragment … our sense of reality' (Gomel 2014: 26). She notes that, under these circumstances, realism becomes inadequate. Science fiction literature prepares us for the moment when we become others to ourselves (Gomel 2014: 27) – transhuman or posthuman.

Her positive prescriptions then are these (Gomel 2014: 27–28). She wants to identify narrative strategies for representing the alien, an 'ethics of metamorphosis'. This takes the alien 'not so much as a positive entity but as an invitation to a change of self'. And Gomel links this attitude to 'the mystical notion of transcendence, in which the numinous is an opening to what lies beyond and outside the space of humanity'. She is therefore seeking not an abstract, but an earthed, metaphysics; she seeks to explore 'the unprecedented, the encounter, and the new, as it takes shape here and now': an ethics of ontological transformation as an alternative to the humanist ethics of the Golden Rule. And 'in interacting with the nonhuman, we become posthuman' (Gomel 2014: 28). Posthumanism is a willingness to abandon both an identity and an ideology. 'Fictional aliens are … the shorthand we use to indicate what lies beyond the self-imposed boundaries of our humanity'.

In practice, as remarked, much science fiction dealing in alien encounters reinforces anthropomorphism, even ethnocentrism. This is true of the stories considered so far, despite their focus on transformation. Gomel points to reactive narrative strategies that borrow from 'psychological realism', such as first-person narration, a sympathetic human protagonist, and focussing on an alien character in a way that humanizes it, techniques found even in the ablest writers. She looks then for 'a residue of alterity that warps the fabric of the text, creating shifts and lacunae through which ... [something] nonhuman seeps into humanity' (Gomel 2014: 28). Although she begins with examples of confrontation and assimilation between humans and aliens, she then turns to instances of transformation, of cognitive estrangement in its purer forms, looking at encounters which alter humans. She identifies three potential axes of transformation: worship, language, and self-consciousness. And she concludes by identifying Lem's stories as offering the best instances of aliens that resist Theory of Mind, empathy and sameness. In short, Gomel offers a thesis about the place of a theological model in science fiction (apophaticism, as opposed to anthropomorphism), a method which allows a focus on difference rather than identity as the basis of intelligibility, and her work begins and ends with an exposition of Lem's ideas.

A final point: Gomel's consistent pursuit of anti-humanist themes presents us with a conundrum concerning positive forms of engagement. Critics tend to agree at present on the earlier materials, identifying imperial and colonial assumptions in the early space operas, the effects of the Cold War in the 'hard' science fiction of the Golden Age, with its emphasis on militarism and the colonization of other planets, and reactions to these themes and the threat of nuclear annihilation in the apocalyptic dystopias produced in the 1950s and 1960s. But if we then criticize any hopes of progress whatsoever – whether feminist, post-colonial, ecological, or queer – as repeating earlier tropes in new forms, what is left?

In his *History of Science Fiction*, Roberts raises a version of the same question, asking 'whether, by creating a new discursive space for the sublime not predicated upon the older religious paradigms but upon a secularized logic of mobilized, systematized materiality, SF necessarily aligns itself with fascism and other such totalitarianisms' (Roberts 2016: 227)? He points to a series of persistent icons in science fiction which could support this view, the obsession with machine technology, the assumption of progress and love of the new, the ever-present desire for saviour-figures and, most of all, 'the bias towards [a] totalising and systematising apprehension of the world' (Roberts 2016: 228). He resolves the dilemma in his own fashion, appealing to an earlier origin of science fiction in the Protestant Reformation, seeing it not solely as a product of the 'nearly vertical' progress of technological and cultural change in the twentieth century. The genre therefore has other resources.

Gomel, however, points in another direction, which is simply to move to a smaller scale and to eschew the search for totalizing solutions, in science fiction as elsewhere. She has identified a number of motifs that we should look out for when considering instances of interplanetary communication, both fiction and non-fiction, looking in particular for signs of transformation on the human side and noting the temptation to introduce anthropomorphic representations that obscure the real story.

There is a particular reason that supports this detailed approach that, to an extent, ignores the overall or stated narrative of the story. For science fiction responds to the visual element that comes with film. Science fiction as a modern genre is unimaginable without the experience of cinema, and science fiction novels are marked by this experience in that they constantly refer to what is seen as the predominant means of telling their story; the story is told through attention to telling detail. The specificity of film as a medium is based in its ability to 'record and reveal physical reality', to gain insight into 'life' through showing things normally unremarked, 'the quotidian and the marginal, the fortuitous and the ephemeral' (Hansen 1997: ix). This quality is taken over into science fiction, and is quite different to the literary novel, where fortuitous detail is rarely part of narrative technique and the visual element generally less prominent. Gomel is right, then, to look to detail and to narrative anomalies to see the sense of what is being conveyed, and I shall follow her in this approach in the next discussions.

IV. Extra-terrestrial languages

These questions of collaboration, evoked at the end of the discussion of Delaney and Stephenson, are important when we turn to real-life projects intending to discern communication from other civilizations and, potentially, to make contact. These projects have to confront the full implications of the model of communication that places all its stakes on the transfer of information between minds. 'Martian' languages – means of communicating with beings from extra-terrestrial civilizations – are marked in science fiction accounts by, on the one hand, a desire for bodiless, mind-to-mind contact, so that information can be conveyed without noise or interference, and, on the other hand, by the problem of intercomprehension: how do we find a common basis to the parties involved so that each may make sense of the other in some compatible way? The first can be stretched without much effort to include supernormal mental powers – telepathy, knowledge of another's thought, clairvoyance, insight into the future – which are symptoms of the desire for contact with another mind. The second, in such accounts, can be 'resolved' either by recourse to a common origin or to comparable experience of practical constraint, which is usually recounted in terms of a common evolutionary trajectory.[9] How do these conditions work out in the search for extra-terrestrial intelligence? It is clear that the project of communicating with extraplanetary civilizations lies at their heart, yet the identity attributed to the 'other' – the correspondent – varies, sometimes appearing to be no more than a way of talking to ourselves.

---

9   In practice, the two accounts play into one another. The principle of a common origin is often supplemented by a concept of time travel, so that citizens from a future state return to our time to encounter and aid their ancestors, and this supplement in turn relies on the vision of a single evolutionary path, where common inorganic materials and invariant physical laws are held to underwrite a single path to the development of intelligence, culture, and technology. Common ancestry and facing common problems mesh. It would be possible to construct a classification of science fiction encounter stories around these variables as axes, and to identify outliers.

Early anticipations of signalling set most of the terms explored by the first constructed language, Freudenthal's *Lincos*, published in 1960, on which later models build. There is, in parallel, a modest history of signalling, with a variety of modes of communication used. The purpose of this section is to trace these two strands and to chart their presuppositions.[10]

An early SETI language

The basic condition for such research lies in the possibility of detecting and sending radio signals: we can search for coded information because we can detect radio broadcasts and analyse these signals using computers (see McConnell 2001: 3). We are looking for evidence of intentional signals, which must be distinguished from naturally produced emissions, defined as noise; making the distinction can be controversial. The question of languages comes up once we have decided not simply to look for signals but to ask what they might mean, and the way chosen to think about this second question has been to reverse the perspective and to ask how we might communicate with other worlds: to define what kind of message we can transmit that might be received elsewhere. The detection of signals also implies the possibility of sending messages.

Early projects imagined signalling to the Moon or to Mars by visual means, creating large scale earth-bound constructions or using mirrors or lights. However, 'Marconi's pioneering work on radio communication [in the early twentieth century] promised a more effective way of communicating across the cosmos' (Oberhaus 2019: 8–9). Once terrestrial radio communication was established as a means, one could imagine both detecting incoming signals from other planets and sending messages. Radiation coming from the Milky Way was first observed in 1932 (Oberhaus 2019: 12), giving substance to the question, if we detected a purposeful transmission, how could we exchange ideas with beings from other planets once contact had been established?

---

10   I have relied for this discussion on the papers collected in Regis (1985), on McConnell (2001) and DeVito (2014) and, above all, on Oberhaus (2019).

The possibility had been discussed speculatively earlier and some ground rules laid down. Francis Galton, inspired by Schiaparelli's detection of the presence of canals on the surface of Mars, had in 1896 designed 'a theory of extraterrestrial message construction based on Morse code' (Oberhaus 2019: 7). Galton's achievement was that he recognized that, given the impossibility of feedback, signals would have to be self-interpreting, so that they would begin by establishing the rules for their being read, such as 'start of message', the notion of 'identity', and 'arithmetical operators' (Oberhaus 2019: 8). Galton also suggested that the material chosen to establish contact would consist in the rehearsal of information such as astronomical facts – the 'principal planets … their … distance from the sun, rotation period, and circumference'. On this foundation, compound signs could then be established to convey more complex concepts. These assumptions of binary signals establishing basic instructions, capable of expressing a logic of identity and contradiction, which would then allow compounds of established terms, represent the starting point for later elaborations.

In 1960, there was a coincidence of two independent projects, symptoms of a wider interest. The first was Drake's Project Ozma, 'the first microwave survey of a nearby stellar system for signs of intelligent life', which we have met; the second was the publication of the 'first developed symbolic communication system for interstellar communication' (Oberhaus 2019: 13). This was Hans Freudenthal's *Lincos* – a contraction of *lingua cosmica* – subtitled (with a hint of Swedenborg) 'Design of a Language for Cosmic Intercourse'.[11]

Freudenthal, whose background was in teaching mathematics, began with basic numerals and built up a language that first described concepts from the physical sciences, but then also more abstract concepts (Oberhaus 2019: 14). Rather than present a 'finished' language, he sought to illustrate the process of learning, so that the recipient might, in deciphering the message, come to understand the sender's world view.

---

11   I draw on Oberhaus's account (Oberhaus 2019: 193–202) of Freudenthal (1960) (see also extracts in Regis 1985: 215–228).

His model replicates many features of human experience. For example, he draws on the idea of phonetic oppositions, distinguishing distinctive units by using different radio wave lengths and durations. The task was then to create a self-interpreting system that conveyed its rules and meanings unambiguously.

A second feature was to assume parallel practices between human and extra-terrestrial language learning. In teaching young children, a good part of instruction consists in showing and telling. Given the separation of the parties, showing is practically impossible, and Freudenthal relied for the most part on the advanced educational practice of getting the 'penny to drop', as the recipients gain insight and learn to manipulate the concepts being conveyed on their own account. Nevertheless, we have to assume the alien partners have a formation like ours, beginning with showing and telling before leading to more advanced learning. Freudenthal drew on this process by beginning the messages with numerals, giving a number of peeps that correspond to a given symbol. On the basis of a number system based in tens, the sender can then introduce some basic operators – addition, subtraction, and identity (equals). The next move is to replace these 'pictures' of numbers with their binary equivalents, for practical reasons of transmission.

So far, the theory repeats Galton's earlier telegraphic system, introducing numerals, basic functions, and binary representation. But we might notice some of the assumptions that have been introduced. First, there is a tension acknowledged between the situated and embodied nature of language learning and the necessity of exchanging disembodied meanings and communicating mind-to-mind. Then, there is a model of the evolution of mind, from a primitive stage where the symbol is the thing to a more advanced stage where abstraction predominates, and so from a world dominated by physicality and emotion to one where reason and intellect pertain. Third, we take it as read that this development is repeated in the child, whose mind is taken to be an intellectual blank slate. And last, we suppose that the recipients of the broadcast – another planetary civilization – will share in the same pattern; we suppose minds similar to our own. These minds could then be like ours but ahead of ours, both in technology and morality, more intellectual and more reasonable, with an advanced culture, having resolved the problems we face of ignorance, poverty, crime, and warfare and, moreover, potentially interested in this world with its enduring problems. While Freudenthal nowhere touched directly on these matters, the presuppositions allow for such developments.

These possibilities are part of the basic picture, shared by Galton and by other early sketches.[12] We shall not follow Freudenthal's both ingenious and charitable constructions as he developed the logical and mathematical concepts and then the 'concrete applications' of this symbolic logic, except to note that other planetary civilizations are thought to share our basic categories of time and logic and that, more generally, a mark of such 'Martian' languages is that they are constructed using what we regard as elementary mathematics, long established and the basis of the education we ourselves have undergone. Perhaps symptomatic of these choices, the system omits any discussion of calculus; giving a description of calculus in *Lincos* would have been a major task, although Freudenthal recognized the oddity of creating a symbolic logic for the purposes of establishing interstellar contact that avoids all the physics that makes such communication possible.

I call these applications 'charitable' because Freudenthal's overall purpose was to convey an understanding of human intention and meaning making and to communicate facts about human behaviour. To this end, instead of giving general examples (as in the case of mathematical and time-related concepts), he sought to 'show' aspects of human life through what Oberhaus calls 'morality plays', instances where two or more humans 'discuss some event that corresponds to whatever behavior is under consideration', for example, matters of love, death, and individuality. Freudenthal anticipated that the extra-terrestrial would be able to generalize further about human behaviour on the basis of these individual examples (see Oberhaus 2019: 199–200). This epitomizes his humanistic, pedagogical approach; he did not define 'right' behaviour but left the recipients to work things out for themselves. He added in a series of new terms, including symbols for true and untrue statements and for good and bad actions; the latter evaluations were introduced as judgements in the morality play, as actors' understandings rather than abstract values.

12   such as Hogben's *Astroglossa* (1950).

Lincos is suspended between Freudenthal's admiration for the ability of natural languages to communicate meaningfully without relying on rigid formalisms (Oberhaus 2019: 201) and the necessity of constructing a formal, logical language. The author claimed, however, not to have followed any particular philosophy of language in constructing his symbolic language but dealt only with 'language as a means of communication' (Oberhaus 2019: 202). As we have seen, in practice he introduced recognizable and well-established models both of how language works and of the beings with whom we may have business. We continue to operate with mind-to-mind communication as the horizon of ambition and with an evolutionary scheme common to all parties allowing the sharing of meaning.[13]

The first signals

It is worth remarking that the first gesture towards signalling consisted not in broadcasts but in small plaques attached to the Pioneer probes, launched in 1972 and 1973 and surveying Jupiter before travelling beyond the solar system; the plaques 'depicted a nude man and woman, as well as a map of our solar system to show the origin of the spacecraft' (Oberhaus 2019: 15). These were followed by more extensive records, both visual and audio, 'designed as a sort of time capsule to represent life on Earth', which were included in the two Voyager missions, which were launched in 1977 and flew past the planets of the solar system and into the beyond. In Sagan's words, these plaques represented an 'opportunity to view our planet, our species, and our civilization, as a whole, and to imagine the moment of contact with some other planet, species, and civilization'

---

13   We might notice a formal parallel with Korzybski's project. Both wished to create a logic-based language whilst remedying the limitations created by propositional and predicate logics, by recourse to human enquiry and open-ended learning in Freudenthal's case, while Korzybski proposed an escape by turning to a non-Aristotelean logic that allowed the mapping of a non-Euclidean, non-Newtonian world. Neither had the technical means to hand that Ollongren later exploits (below).

(Sagan et al. 1978, cited Oberhaus 2019: 15). These gestures cannot be seen as strong attempts at communication other than with ourselves.

In terms of a practical attempt at signalling, only one effort was made at messaging to extra-terrestrial intelligences before the cutting of public funds for SETI in the early 1990s. In 1974, Drake and Sagan 'sent the first interstellar message from the Arecibo telescope in Puerto Rico to a star cluster approximately 22,000 light years from Earth' (Oberhaus 2019: 14). However, *Lincos* was not employed in this attempt, so there is a second tradition to be considered. The investigators used a 'simpler unambiguous method' than *Lincos* (Sagan's words), modelled on a prototype message Drake had designed in 1961 after the initial conference on searching for extra-terrestrial intelligence at Green Bank, West Virginia. This message 'consisted of 1,679 binary digits arranged as a rectangular bitmap. The resulting image depicts the numbers one through ten and the atomic numbers for the five elements that make up DNA, as well as the formulas for the sugars and bases in DNA nucleotides, a crude drawing of a human, a graphic representation of the solar system, and a picture of the Arecibo telescope' (Oberhaus 2019: 14).

We might notice the reliance on numerals and the pictorial nature of the representations using a binary code, as well as the contemporary focus on the genetic code as a clue to human identity (see Oberhaus 2019: 171–177). Oberhaus notes that, although designed for 'intelligibility and transmission', the messaging was for the most part a demonstration of the 'capacity of the new instruments on the Arecibo telescope' rather than an attempt at interstellar communication (Oberhaus 2019: 14). In this, it resembled the plaques on Pioneer and the records on Voyager. It would in any case have taken 'approximately 50,000 years from the time the message was sent to receive a reply' (Oberhaus 2019: 171).

Later signals and SETI languages

Oberhaus lists a small number of later messages. In 1999, two messages 'inspired by Freudenthal's *Lincos*' (Oberhaus 2019: 16), building from an introduction to arithmetic to more complex topics, were transmitted from the Evpatoria radio telescope in Ukraine. The 'Cosmic Call' messages 'used a unique symbol system to convey a wide variety of topics ranging from basic arithmetic to the physical composition of Earth's crust'. It also included the original Arecibo message (for details of the overall design and content, see Oberhaus 2019: 179–191). The two identical broadcasts were aimed at 'a star 70.5 light years away in the constellation Cygnus' and at 'three other stars located between 51.8 and 68.3 light years from Earth'.

A second Cosmic Call emission was made in 2003, also from the Evpatoria telescope, repeating the content of the first together with the addition of the binary code for an elementary chatbot, Ella. Provided the recipients could run the software, 'Ella represented a … creative departure from typical interstellar message design and prefigured message designs based on artificial intelligence' (Oberhaus 2019: 17). In 2001, from the same source, 'the first musical interstellar message was sent … in the form of a theremin "concert for aliens"' (Oberhaus 2019: 16).[14]

In terms of symbolic systems, there are two innovations in the recent period. Both systems start from Freudenthal's *Lincos*. Carl DeVito and Richard Oehrle published research in 1990 'on a novel communication system … based on the "fundamental facts of science", which demonstrated that it was possible to build a communication system capable of conveying a great deal of our scientific knowledge using basic arithmetic as a foundation' (Oberhaus 2019: 15–16). This was an attempt to tidy up Freudenthal's humanistic impulses and concentrate on a scientific language based on literal correspondences. And the mathematician Alexander Ollongren published *Astrolinguistics* in 2013; in Oberhaus' summary, 'the text is essentially a second-generation version of Freudenthal's *Lincos* that applies recent developments in computer science to the problems of interstellar communication' (Oberhaus 2019: 17). The key innovation was to include a translation function to interpret the message at a second level of the signal. Despite the seeming optimism contained in Freudenthal's original proposal concerning the potential for communication between minds at a distance, the later uses of the model are openly sceptical about

---

14 Echoing the alien musical messaging anticipated in Mary Doria Russell's novel, *The Sparrow* (1996).

this possibility, and the justification offered for such projects is placed increasingly within our world.

DeVito and Oehrle[15] suggest that, against Freudenthal's minimal assumption that aliens can reason as humans do, which then demands a great deal be transmitted about the language before worthwhile information can be conveyed, a good deal of common scientific understanding can be assumed because of the shared means of communication. In DeVito's account, 'it is assumed ... that our correspondents can count, understand chemical elements, are familiar with the melting and boiling behavior of a pure substance, and understand the properties of the gaseous state. All this should be known to any society capable of developing the radio telescope' (DeVito 2014: 185). No matter what the cultural differences between the two civilizations, there is a reality which can be shared when describing 'the same phenomena, particularly uncomplicated chemical or physical processes' (DeVito 2014: 187), and 'this additional assumption [to Freudenthal's minimum] enables us to move more rapidly to the point where interesting information can be exchanged'.

Both DeVito and Freudenthal are concerned with the construction of a formal logical language without recourse to any natural language; the alien recipients are, as it were, a pretext. Freudenthal used the constraints of a hypothetical alien civilization and the use of radio signals as a means of giving limits to the development of a solution to this problem (DeVito 2014: 106). In similar fashion, DeVito has practical aims within the human sphere; he suggests such a project may allow the development of a computer language which 'incorporates science into its basic structure' (DeVito 2014: 186) and, moreover, that such a language may serve as a basis for machine language acquisition which would in turn aid the development of artificial intelligence. In particular, these achievements would promote the possibility of 'machine translation', translation between natural languages without recourse to native speakers.

The postulate of an alien recipient is then a feature of the model; its function is to eliminate both the arbitrariness of natural languages and all trace of embodiment. We seek a correspondent who is effectively a

15   I rely on DeVito (2014).

bodiless intelligence, with whom we may communicate mind-to-mind.[16] Despite the acknowledged absence of any recipient (through distance), DeVito's practical aim is to reach the point where 'precise, useful information' (DeVito 2014: 188) can be exchanged. By restricting the exchange to the level of information, DeVito loses sight of Freudenthal's humanistic exercise in conveying experience of a human world view, much closer to the practical use of language in the world. Although they both begin with notions of equivalence, based in Freudenthal's case in the 'pictures' of numerals, DeVito sticks to the demand for literal correspondence, believing it to be the basis of scientific knowledge, while Freudenthal suggests a world of metaphor, of speaking of one thing in terms of another.

Yet, although he begins with 'real' objects that can be shared through exchanges of information, when he comes to the arithmetic of complex numbers – important, he notes, for digital signal processing – DeVito raises the question of whether such mathematical objects exist outside our minds. He acknowledges that his view of communication demands the objective existence of objects, and therefore that the whole project may fall if numbers are ideal (see DeVito 2014: 112), and he does not consider the possibility of mixed objects, of things that can claim both existence and non-existence. It is then significant that Ollongren starts with a logic that rejects the law of the excluded middle, which states that a proposition must be either true or false.

Ollongren's project may be seen as fulfilling DeVito's hope of creating a computer language which incorporates scientific logic, for he draws on developments both in digital computing and mathematical logic to design a logical language that will allow an extra-terrestrial intelligence to translate messages sent from Earth in a natural language.[17] The revised language is called *Lincos 2.0*. In distinction to Freudenthal, Ollongren proposes a 'multilevel approach', providing a basic text containing the message (which

---

16 This recalls the conditions of the Turing test: despite the purpose of the original game, which was to determine the gender of the hidden correspondent, the test eliminates all consideration of the body and so can be used to define machine intelligence (Descombes 2001: 110–115).
17 In giving this account, I rely solely on Oberhaus (2019: 202–223).

can be made up of text, images, and other media) together with a second metalevel to the message which will 'provide a means of understanding and interpreting the ground text' (Oberhaus 2019: 213). This second level must be self-interpreting with simple syntax and rich content, though Ollongren adds other criteria – linear notation, redundancy, and the possibility of structuring and sizing the messages. In short, 'this metalevel is essentially an annotation of a text in natural language that is coded using the calculus of constructions with induction' (Oberhaus 2019: 213–214).

We shall not investigate the calculus of constructions, except to say that it allows the creation of a language capable of realizing automatic translation between natural languages. Through the affinity of computer programming and abstract logic, we gain mutual transparency. This again is a project at a human scale. *Lincos 2.0* is conceived as describing the states of affairs created by objects – these are the language's defined terms – and the relationships between facts are articulated through the conventions of the calculus. A fact is structured, and this structure is determined by the proof which is used to create the fact. In this fashion, if you get the logic right, you can produce a self-interpreting formal language.

The alien, then, has to realize that we have utterances and rules which form the utterances. Deciphering will be achieved by use of a computer. The sender transmits a message in a natural language together with a metalevel showing 'the logical form of basic text ... accompanied by its proof' (Oberhaus 2019: 220). The alien 'automatic information-processing machines' will recognize the meta-message uses constructive logic and discern the appropriate signs and functions, allowing the terms common to the two levels to be identified and so enabling a move to the natural language in question. In sum, 'at its core, *Lincos 2.0* is an effort to provide an annotation to a natural language (or visual message) used in interstellar communication. The hope is that by roughly articulating the logical syntax of natural language, this will aid the interpretation of the natural language message' (Oberhaus 2019: 221). Oberhaus adds that, 'of course, the semantic content of this message may be impossible for the extraterrestrial to reach', with the now-familiar thought that, to establish common meanings, 'it will be necessary to start with mathematical or scientific notions that may be presumed to be universal'.

Discussion

Both DeVito and Ollongren prolong the ambition of mind-to-mind communication, to that end focussing on a part of the spectrum of human learning that lends itself to unambiguous facts capable of being transmitted together with knowledge of a limited and fixed context. Confronted with the extraordinary distances between hypothetical correspondents and the immense periods of time any exchange would have to take, languages designed for interstellar communication succumb to the temptation offered by a model of the bodiless transmission of information from mind to mind. The model conjures away the layers of political agreement, negotiation of funding, engineering solutions and collective scientific research, all relying on embodied conceptions, which are required for messaging to take place and local versions of which, one must assume, must also exist elsewhere for there to be any kind of correspondent civilization to engage with.

I have two remarks to make about these artificial languages. The first is to note some sort of paradox. The notion of contact assumes that intelligent life will have built radio-telescopes and developed the ability to analyse data, like us, and seek to communicate with us. They will understand enough physics and mathematics to handle electromagnetic radiation and create an information technology: in essence, they will have invented the equivalent of the gramophone and the typewriter, which lie at the origin of our technologies of radio waves and computers. Yet the language we are trying to develop, and the ways we anticipate their attempts to communicate with us, does not consider the scientific revolution in information handling which has permitted our explorations of space and, presumably, theirs. It has a dream-like aspect, assuming their formation will be in Euclidean geometry and Newtonian physics.

The second comes from the first. The image of a rational language which has been developed from Leibniz and Locke onwards concerns the ambition of unambiguous communication of information between disembodied minds. In practice, as we have seen, the experience of seeking such communication deals largely in its frustration, in seeking to overcome interference, to interpret signal shifts, to distinguish messages from noise, not to mention the problems of transmission over great distances, nor the possibilities of messages being lost, or intercepted by those not meant to receive them. This brings us back to the earlier observation, that the communication of information is not the same thing as speaking a language, and one should not be taken for the other.

A focus on information therefore repeatedly brings intimations of the greater difficulties and limits imposed by our reliance on language and even non-linguistic forms in our everyday lives. In this fashion, the theoretical projects threaten to repeat the practical encounters and frustrations met with by earlier attempts to realize interstellar communication in a variety of settings. The high level of intelligence at work in each does not overcome the ambiguities in the presuppositions of what communication is and does. We might conclude that the search for extra-terrestrial intelligence has its equivalent period to the Golden Age of science fiction, a period when communication between planetary races can be conceived in an optimistic, unproblematic perspective and hopes have not yet been dashed by experience, and that these extra-terrestrial languages originate in this period.

V. Making connection

We may conclude that a number of ideas are shared by the fictional and the real language projects. Both are framed by the desire for mind-to-mind contact, on the one hand, and the problems of inter-comprehension, on the other: communication independent of bodies, conceived as information, and overcoming the obstacles to this ambition; in a phrase, the hope of telepathy balanced against the fear of solipsism. In his survey, Oberhaus has to take the hope for granted, although he repeatedly mentions the technical problems, the nature of the physical interference to be overcome, the distances involved and the time lag they imply, the improbability of messages sent being received and understood, the impossibility of establishing exchanges of information or a conversation: together, these constitute a formidable array of obstacles. He concentrates instead on the second topic, how a common understanding might be developed once contact at a distance has been established. Here the question comes down to how to anticipate a common ground which would allow minds to touch, an explanation based in evolutionary responses to constraints, with either distant shared ancestry or convergence allowing – or disallowing – communication.

We have noted some themes which emerged in the descriptions of the various artificial languages. Freudenthal assumes a model of progress, applied to children and humankind alike, from concrete pictures to abstract, manipulable representations, which aliens may share. Linear time is an expression of this evolutionary model. Atomism is important to the development of the models, with atomic facts arranged in states of affairs, with relations and relations of relations being mapped. In this fashion, we build from the single mind to a shared totality. This approach is tied to the assumption that arithmetic and basic science provide the basis for intelligible communication and a logical system of signs. And the language constructed is approached through fixed rules and utterances; ambiguity, improvisation, creativity and intelligence are at best deferred to compound signs or higher metalevels.

Oberhaus offers sophisticated commentary on these themes and engaging with that commentary will allow us to fill out these conclusions. In particular, because he explores the presuppositions that permit these artificial languages to confront their tasks, it allows us to make connections to the earlier discussion of models of language.

Presuppositions of the model

At the outset, Oberhaus notes that the artificial languages are caught between the options of a serious attempt to communicate with extra-terrestrial intelligences and positing alien recipients as a fiction, allowing the design of translation programmes, for example. If we take the first position as a thought experiment and confront the task of messaging extra-terrestrial intelligences – taking for granted the difference both in the energy consumption needed and the possible coverage between listening for signals (SETI) and sending messages (METI: Messaging

Extra-Terrestrial Intelligence) – the focus is clearly on conveying human thought intelligibly to an alien subject or, better, exchanging thoughts. With regard to the discussion (above) between particularist and universalist models of language, Oberhaus is clear that languages for signalling at a great distance of necessity take the Chomskyan side of the argument concerning intercomprehension. Chomsky posits an innate human biological capacity, a 'recursive generative procedure' (Oberhaus 2019: 29), so that each of us is born with a faculty that he calls 'universal grammar'. In this fashion, we use 'finite means to express an unlimited array of thought' (Oberhaus 2019: 28), overcoming the limits of the 'tabula rasa' model, whereby a child learns to manipulate a limited body of taught examples, and allowing the possibility of translation, overcoming any 'indeterminacy of reference' inherent in demonstration or pointing. In short, the structure of the human brain allows language acquisition.

The constraints of the programme imply that (quoting Chomsky) ' "If a Martian landed from outer space and spoke a language that violated universal grammar, we should simply not be able to learn that language the way we learn a human language". Instead, we would have to "approach the alien's language slowly and laboriously – the way that scientists study physics, where it takes generation after generation to gain new understanding and to make significant progress" ' (Oberhaus 2019: 31).[18] In order to further the possibility of translation, we would also need to explore the differences between our two species in terms of their neurobiology, cognitive modules, and sensorimotor systems.

While compatibility appears on the face of it improbable, nevertheless Oberhaus argues there are grounds for supposing aliens might think like us. He appeals to an argument presented by Marvin Minsky (1985a, 1985b), who began by proposing two constraints to which both parties respond, on the one hand, the laws of physics, uniform throughout the Universe,

---

18   We might notice that, although the image is a striking one, the implication is that both physics and language have given, stable objects, conceived to stand completely independent of their human investigators, of which one can advance one's understanding progressively by effort. This is not a realistic vision of the interpenetration of the human and natural worlds, nor of intercourse with other worlds.

and, on the other, that 'all life is subject to resource scarcity' (Oberhaus 2019: 32). Technological progress, central to the possibility under discussion, is (in this argument) driven by responding to resource depletion. From this starting point, one can define intelligence in terms of its independence from the specific nature of human intellectual activity, determined instead by the nature of the problem in hand. 'If all intelligent creatures are faced with the same fundamental problems (i.e., restraints on space, time, and materials) and the methods of intelligence are determined by the nature of the problem at hand, Minsky reasoned that extra-terrestrial intelligences will arrive at solutions similar to our own, namely symbolic systems for representing these problems and processes for manipulating those systems that can also be described symbolically' (Oberhaus 2019: 33).

In addition to scarcity and objective intellectual processes, Minsky added a third principle, that of 'sparseness': alien symbolic systems will not differ from ours because every evolving intelligence will come upon the same ideas since 'they are simpler than other ideas that result in the same process' (Oberhaus 2019: 34); numbers and arithmetic offer examples, which is why they may serve as a good basis for interstellar messaging.

This messaging model has a clear set of components. It deals in messages between minds; minds are located in individual brains, with their cognitive possibilities and constraints; and we can universalize this model on the basis of need, problem solving, and Occam's razor.

We may make three observations. Minsky, and Oberhaus after him, in proposing a model based on scarcity, problem solving, and sparseness, presume a human-shaped figure in the argument: in order to solve problems of resources efficiently you require a stomach, a head, and limbs, responding to biological need, with thought to identify your ends, followed by action. At the very least, a human figure underwrites the argument.

Then, on this account, Man is an intelligent *animal*; it is, at the least, a reductive model, and it may miss out everything that distinguishes the human from the rest of nature.[19] Animal behaviour, we may suppose,

---

19   This illustrates the argument of how the concept of information situates the 'problem of man' in a far wider framework than has been traditional and may, by doing so, miss several vital features.

is strictly functional: it deals in matters like food, territory, mate selection and other needs, and it is confined to the immediate environment, to the present both in time and place. Yet humans, although an animal species produced by the same evolutionary principles, have made a step-change: they have moved beyond the primacy of need and functional responses and are not confined to the here-and-now: instead, they are capable of imagining different times and places and for these acts of imagination to alter their behaviour decisively in the present. The uses their biological bodies are put to, even meeting their functions and needs, are in the human case determined by the invisible properties of the mind, by things that are not present. Animals, we might say, are defined by needs and by the present, while humans are shaped by the future and the past, by desires and by obligations: what might be, or what ought to be, or what must be, the case.

The distinction animal mind/human mind is therefore shaped by a series of oppositions, the one present, the other stretched between past and future, the one concerned with will, the other with memory and hope, the one with biological needs, the other with human desires (which may include some very strange developments in terms of putting biological needs to work), and so forth. In short, attempts to describe human behaviour as being in continuity with animal behaviour will miss what is essential about human beings (cf. Conway Morris 2022).

Third, this distinction points to the absence of any sociological dimension in the cognitivist approach, and the assumption that human minds are indeed animal minds in their crucial aspects reinforces the model of language which confines its vital properties to the individual human brain. Before asking whether we can assume similar enough conditions – of interior thought, a shared universal grammar, and parallel conditioning bodily and brain structures – to allow alien life to comprehend our judgements, we might comment how much has been left out concerning human life. From the perspective of a social scientist, this combination of individual experience and universal protocols ignores the intermediate scale on which social action is effective; it has no account of collective life and history. The challenge for a cognitivist approach, in this optic, is to account for the shared ordering of human desires, ideas and habits, and the scale on which they occur, their independence from individual needs and instincts, and the wide range of cultural variation in comparison with the limited range of human emotions. A sociological approach to another culture demands a different understanding of the object, language, explanation, comparison, and time. In effect, in the cognitivist position the evolved structure of the brain is required to do the work that falls in practice to various human communities, operating at different scales and over varying periods as they develop collective representations.

The counterargument to this sociological claim – that language is used as a means of pursuing collective projects, that it is a tool but may not be the defining feature of the human species – would be that this approach misses the force of the position that the primary use of language is 'internally to order thoughts and only relatively infrequently to convey these thoughts to others' (Oberhaus 2019: 41). We may identify this latter position loosely with that of Locke: a man's thoughts are his property, and we are then left with the problem of how private thoughts may be conveyed between people. Interstellar communication takes this problem to its extreme point.

Minsky believed it possible to derive the characteristics of human language without recourse to the social environment. He begins with the representations of 'things', allowing the development of more abstract notions of properties and then relations between these properties, dividing the world into significant relations, which in turn permits the concept of 'cause'. This use of representations allows the crucial feature of human intelligence and language as a means of communicating that intelligence, whereby the idea of a 'thing' enables a speaker to embed one clause within another, permitting the representation of 'prior thoughts as objects in another thought' (Oberhaus 2019: 35). In this fashion, what starts as skills (mastery of causes) can be employed in general intelligence, in which symbols are turned on themselves (see Minsky 1985b: 123ff.).

Minsky claims that these processes, born of the constraints imposed by the laws of physics and the scarcity of resources, can stand independent of realized mental processes, whether human or extra-terrestrial, and allow us to suppose interstellar communication would be possible. And he supports this claim with the analogy of computer languages: we already have evidence of precisely this kind of intelligence distinct from human embodiment, in Artificial Intelligence. Yet this argument takes for granted the immense collective human labour needed to construct machines capable of communication in this sense.

In sum

Given this Lockean frame, Oberhaus's review of the various designs for interstellar communication identifies three tendencies. In the first place, they tend to focus on the task of recognizing structured – and therefore intentional – information in potential signals. Then, despite promises, they bracket off the possibility of conveying new understanding, as opposed to initial exchanges of known information. And last, they increasingly focus on the value of the work as lying within the human sphere and retreat from any focus on contact.

Beyond the role of language, a second theme in the search for a possible basis for interstellar communication has been the potential of mathematics to provide a universal language. I shall not pursue this argument but offer two observations. First, we need to pose a version of the earlier question: does communication between human groups start from counting and proceed to comparing accounts of celestial events? Almost certainly not. Although these topics are believed more or less universal to humans, groups in practice communicate by giving and receiving gifts, including people – wives, children – and services of various kinds. Counting (and equivalence) and understandings of other kinds are built into these exchanges and serve them, they do not precede and shape them. There is then a strong assumption of Western presuppositions and economic rationality as we offer codes to extra-terrestrials and seek to effect transactions. We assume a trade in ideas, in intellectual property. And what might we get in return? For the breakdown in relations, in normal human practice, is often tantamount to war, not simply a missed opportunity to exchange goods.

And second, we may remark that, perhaps counter-intuitively, Oberhaus resolves the problem of intercommunication by positing problem-solving physical beings at both ends of the line. He argues that embodied cognition is the key to found relationships between the primitive elements of thought, and so, bodiless communication can only be conceived between incarnated creatures. The dilemmas noted earlier, which are produced by ignoring the work of social relations, reappear at this point. For we have some idea about the scales at which embodied creatures interact; they comprise encounters between shared systems of classifications, collective representations borne by and negotiated between small groups of intelligent agents and expressed in interactions that are particular both to place and time, in acts of mutual interpretation embodied in social events. In short, mind is found outside the individual brains, and explanation will be embodied not in principles but in descriptions of particular histories, each a nested set of contacts and events, in which mental events are constructed through locally shared narratives. Embodied cognition is best conceived not as the shaping of individual perceptions by the constraints of bodily architecture, but as the production of complex social phenomena, for example, the safe bringing into port of a large naval vessel (Hutchins 1995). In this perspective, intellectual projection can take us only so far; we have to await real encounters to provide us with appropriate materials.

Oberhaus then knows what he wants but mistakes the appropriate intermediate forms of life to focus on. He attributes the labour in practice carried out by social interaction and history to the individual mind, on the one hand, and the Universe in the form of the shape of cosmic evolution, on the other. The one is on too small, the other too large, a scale.

For this reason, he ends by defending messaging extra-terrestrial intelligences within a solipsistic horizon. He charts the arguments made against METI: that we may draw hostile attention to ourselves, that a project that has yielded no positive evidence over so many years may count as unscientific, that it wastes resources better spent on more assured research outcomes, and that it is wrong to leave the 'diplomatic' representation of the Earth to other worlds to a narrow group of interested scientists, engineers and entrepreneurs (see Oberhaus 2019: 155ff.). The first and last objections, of course, fall within an 'optimistic' perspective: they assume there may well be other civilizations with which communication is possible. The other two are more ambiguous and come down to arguments about the deployment of resources; there may be better ways to conduct the search.

Yet, after his thorough survey, Oberhaus concludes the value of the projects considered can be justified without reference to any version of the extra-terrestrial hypothesis. In short, the projects encourage human self-reflection, promote public engagement with the space programme, and are a means of advancing message design and machine learning (see Oberhaus 2019: 164). We are confined within the human sphere; even if we begin by imagining we might receive and read the constructions of other civilizations and dream that we might reply, we end by talking only to and about ourselves.

CHAPTER 4

Four novels of 'first contact'

The history we have traced of the search for extra-terrestrial intelligence and, indeed, projects concerned with messaging other civilizations, even if small-scale in practice, has been accompanied by a series of science fiction stories that acknowledge and exploit the detail of these intellectual and technical explorations. The interplay between the two 'zones' has been considerable. 'First contact' between human and other civilizations has been a staple of science fiction writing since *The War of the Worlds* (1897),[1] and there are many novels that take the notion of contact to frame the exploration of different narrative or psychological dilemmas, drawing on a wide range of human experiences to provide analogies, for example, colonial encounters, missionary work, studies of animal behaviour (particularly of social insects), diplomacy, and warfare. My concern now, as a conclusion to this essay with its focus on technical means of interstellar communication, is with four stories that deal more narrowly with the technology developed and the social setting of the appropriate research institutions: a sub-set of the genre. Together, they track a literary history framed by an evolving technical and political setting.

Taken as a sample, these stories map the classification of possibilities that has emerged in the various histories outlined so far in the essay, as well as offering a variety of comments on the potential for communication with other planetary civilizations. Given that all the writers explore earlier resources (as Heinlein uses pre-War models), they cover the entire period, and we may divide the stories into two types in their judgement

---

[1] The term 'first contact' seems to originate with a 1945 short story of that title by Murray Leinster.

of the present: broadly, optimistic and pessimistic accounts, split between the hope of direct mind-to-mind contact and the less sanguine realization that we may never move beyond the human sphere and that we talk only to ourselves. These two accounts in fact display distinct attitudes; the first manifesting the belief that humans by intentional action can solve problems and achieve predictable ends, the second focussing on the conflicting and, indeed, contradictory interpretations at work in every context, leading to a more questioning attitude of detailed observation and description. Particularly the second may be accompanied by hints of a third possibility, that maybe encounter would involve something that has been excluded by the concept of communication which has come to organize our thought and research, in short, pointing to a shift in perspective. This possibility, however, is not our main concern.

I. Early materials

The first novel is James Gunn's *The Listeners*, published in 1972. I place Gunn first because he gives us samples of the source materials he used in sections between the chapters (which were originally published as stand-alone pieces in science fiction journals), making clear he is writing about the early period. A key source, alongside Morrison and Cocconi, Drake, Sagan and others, is the Brookings Institution's 1960 report 'Proposed Studies on the Implications of Peaceful Space Activities for Human Affairs', written for the Committee on Long-Range Studies of the newly-formed NASA.[2] After discussing a range of issues, including satellite-based communications systems, 'weather predicting' systems, space industries, and implications for foreign policy, the report reviews potential reactions to the development of space flight, including a section on 'the implications of a discovery of extraterrestrial life'. This section acknowledges the likelihood that there may be 'intelligent life in many other solar systems'.

2   Available on the brookings.edu website.

Over the next twenty years, the report foresees three possibilities: such life may visit the Earth, if they have the appropriate technology; human exploration of the Moon, Mars or Venus may reveal artefacts left behind by these visitors; and they may make contact by radio; the section indeed begins with 'efforts to detect extraterrestrial messages via radio telescopes'. Having set these premises, the report then focusses on possible human responses to such contact, distinguishing between the roles of national leaders and the wider population. In the case of the wider population, anticipated reactions include possible greater human solidarity in the face of an intelligent, non-human species, or (on anthropological evidence of confrontations between unequal civilizations) the dual possibilities of cultural disintegration or, alternatively, the adoption of superior alien values, or, again, given the time lag between exchanges, popular indifference. The report recommends further study of the potential intellectual and emotional responses to the discovery of intelligent extra-terrestrial life, as well as studies of past reactions of both populations and leaders to 'dramatic and unfamiliar events or social pressures'. It focusses particularly on the respective roles of scientists and political leaders in such circumstances and the need to control and present information to the public.

This report sets the parameters for Gunn's novel. It is set fifty years in the future, describing a government-funded listening project using the Arecibo telescope in Puerto Rico. After fifty years of waiting, with increasing anxieties about funding and pressure to close down from scientific critics, a message is received; it marks its artificial nature by rebroadcasting fragments from the earliest human radio transmissions in the 1930s. The direction of the signal, combined with the time taken to first receive and then return the broadcasts, allows identification of the sending planet, Capella. If the beacon consists in phrases from early soap operas and comedians, the message is contained in the 'static' between the fragments. It is deciphered, using a computer system which is held to record everything which takes place in the listening programme – a total record – and analyses it for significance. The message follows a type proposed by Drake in 1961 and, when decoded, shows a pictorial representation of a bird-like person.

With the help of journalists they have won to their cause, the scientists overcome the objections of religious fundamentalists ('solitarians') and the hesitations of political leaders, so that the project is continued, and a message sent acknowledging the contact. They also calm public fears and manage expectations: the picture is thought to indicate the Capellan sun is dying and their civilization threatened, and moreover any response will take the best part of a century to arrive. In that time, assisted by population control and extensive social welfare programmes, humankind learns to live together in a sense of unity, peace, and anticipation. When the reply comes, it is sent from an automated system awaiting a signal of acknowledgement; after the transmission of a primer, with a vocabulary, numbers and operators, we learn that the Capellan people have been long destroyed, but the transmitting computer downloads the entire records of their civilization, permitting the peaceful and controlled transformation of human culture. We are their inheritors, benefitting by a posthumous legacy from an advanced culture; it is an example of untrammelled communication, mind-to-mind; a model scientific exchange, if of a one-sided nature.

This is then an optimistic account of perfect bodiless communication: the mechanisms of signalling, reception, deciphering and replying all work like clockwork, the potential human obstacles of funding, political interference, public fears or expectations are managed, the interest of the project preserves human memory and focus over a period when one might have expected priorities to have shifted radically, and the anticipated rewards are detailed, the exchange or, rather, gift, of advanced technical know-how, without a trace of any reciprocal obligation (they are dead) or anticipated threat. One might note that the concept allowing the story is that of total recording, both on the part of the human project's computer system, which enables analysis, and in the notion of unloading the records of an entire civilization, by which it gains some kind of immortality, indeed, elements of which actively enter into the human computer's operating system on the last page. We are nothing if not information.

Gunn gives ideological expression to the self-presentation of the space programme, secure that pure research justifies state funding, and confident in the benignity of all participants. He ignores the defence background, the negotiation of arms treaties and the Vietnam conflict, the consequent mixed motives for funding and participants alike, and as well has no fear of any downside to total recording. In short, he isolates space research from a series of real-world features and tells an unproblematic story.

Anticipations of contact: Solipsism

The next story considers the same world from quite another angle. It was in fact published earlier: Stanislav Lem's *His Master's Voice* was published in Polish in 1968, though translated into English only in 1983. There is a Polish film adaptation. It is a remarkable book that evokes themes from every part of the history we have been considering.[3] It has the same basic plot as the optimistic stories and shares many of the same elements: a message from another civilization is identified, scientists differ as how to decipher it, politicians and security organizations intervene once the broad implications of the discovery become apparent, a project – His Master's Voice – is set up, there is a certain success in deciphering the message, and the Senders remain off-stage throughout. And a good deal of space is given to a variety of theories that seek to account for the portion of the message that has been read, together with speculation on its provenance. All this is familiar territory.

But the tone is quite different and, we might say, deflationary. The narrator, Peter Hogarth, is a mathematician with a critical cast of mind, recruited as such to the 'Project' after a year during which not much progress had been made. Hogarth repeatedly points out the flaws and inconsistencies in the standard narratives of discovery and decipherment and in the presuppositions the participants bring to bear. He begins by casting doubt on the personal motives that animate researchers, illustrated by his own character and history, attacking the motif of the impersonal search for truth. He then draws attention to the rivalries that equally drive and frustrate the project: above all, the clashes between scientific and defence

---

3   It is worth noting two other 'contact' novels by Lem, *Solaris* (1961, English translation 1970, twice filmed, 1972 and 2002) and *Fiasco* (1986, translation 1987). There is a good deal written on Lem; see, for example, Jameson (2005, chapter 8), Gomel (2014, chapter 6) and, for a recent survey, Lethem (2022).

and security priorities, but also the contrast between the free discussions of academics and the priorities and nervousness of administrators put there by the state departments funding the enterprise, the rivalries and incomprehension between natural scientists and humanities scholars and, within the natural sciences, the distinct approaches and tribalism of biologists and physicists. He also identifies the extraordinary sequence of chances and chancers – petty criminals, confidence men, flying saucer enthusiasts and occultists – involved in the initial identification of the code which, although it starts from the observations of neutrino emissions made by two early-career researchers at the Mount Palomar Observatory, soon escapes from any scientific oversight, which is only recovered once Washington takes an interest. By the time scientists were recruited to the project and funding secured, the security interests were in silent but effective control. Moreover, in the aftermath of a 'crisis' in which the rival values of the researchers and security become clear, it appeared that the Pentagon had been running a parallel project, a 'ghost HMV', and that their distrust of the scientists had been a factor from the start. The scientists, then, in contrast to Gunn's positive tale, are displaced from any priority or initiative, both in terms of their internal motivations and of organizational autonomy. They are likened, indeed, by one of their number to domestic animals kept by the state. And the narrator includes a series of insights within this perspective, covering such topics as the creation of interstellar languages, the history of space research, and the search for extra-terrestrial intelligence, as well as analysing the disappointing nature of much science fiction from the scientific perspective; the author displays throughout, in the words of one critic, his 'command of convincing jargon and absurd logic'. Lem's work is perhaps best considered as a satire on the relatively young space programme and the practical and intellectual challenges of working in such a complex environment.

All this is before we reach any consideration of the message – 'the letter from the stars' – or of the Senders. The prospect of 'contact' is in practice an accident that occurs within this complex setting. The story, however, presents two features of particular interest that derive from that accident. In the first place, Lem considers the creative potential of human interaction with the message, which produces new effects in both biology and physics, interactions that are taken to offer potential insights into the Senders and their purposes. In short, he starts from evidence of novelty within the human sphere. In the second place, despite this insight, he – or rather, the narrator, Hogarth – cloaks these discussions in terms that do not differ from those that emerge in the optimistic accounts: the idea of an advanced civilization with benign intentions, concerned with the promotion of life in the Universe. It is true that some of the accounts proposed by physicists in the novel refer to an impersonal framework of physical determinism, but Hogarth dismisses these, pointing to the arbitrary introduction of assumptions to make the reductive arguments work and, at the same time, defending the appearance of intentional aspects in the data. His clinching argument on the last point concerns the 'safety mechanism' included in the instructions and the physics they describe, which effectively prevents humans from developing the destructive powers they have glimpsed into effective weaponry.

Despite this evidence of intentional design, Hogarth's theme is that, confronted with the productions of an advanced civilization, we cannot escape our limitations and our incapacity to understand. He condemns both the aspiration to ready communication and the fear of being alone in the Universe as infantile, but, in practice, we are still unable to transcend our own categories and consequent ignorance.

This solipsism, reinforced by the sceptical character of the narrator, is nevertheless a good deal more sophisticated than a simple refusal of the possibility of knowing for, as remarked, the investigating scientists are able to produce new effects from their engagement with the code (this has something of Delaney's conception of the power of language in *Babel-17*). The code lacks any primer, yet, in a manner not explained, some small part of it (3–4 per cent) has been sufficiently deciphered to allow two research groups to manufacture a product and investigate its properties.

The first, biochemists, produced a gelatinous substance or colloid with certain properties which resemble life: although lacking any metabolism, it responds to stimuli and can produce movement and extension, separating from and reintegrating with the main body; it apparently does this by producing and reabsorbing small amounts of energy in an unstable 'cold' nuclear reaction produced in the separating and joining of large molecules.

The biochemists call their product 'Frog Eggs'. The second group, physicists, handle their material with more respect, keeping it in a reinforced tank and wearing protective clothing when approaching it. The distinctive phenomenon they worked on arose from their observation that flies trapped in a flask, when brought close to the glass surface of the observation window, were first paralysed and then increasingly agitated, suggesting some sort of radiation detected. They called their material 'Lord of the Flies'. The only difference between the two products was the volume of material: 'Lord of the Flies was Frogs Eggs – but in a quantity exceeding two hundred litres' (Lem 1983: 120).

In each case, further investigation revealed new properties. In the biochemists' case, it was found that exposing the colloid to the original neutrino message enhanced the production of complex molecules; the message had the property of favouring the possibility of the formation of life. This is termed a 'biophilic effect'. In this fashion, we may note, we move from the idea of some coded communication, beyond even the notion of instructions or a blueprint, to the possibility of some direct activity in the world, likened to the 'thing itself' by Hogarth, a beneficent effect indifferent to the intelligence or otherwise of the recipient.

In the case of the physicists, a researcher investigating the 'cold' nuclear reactions of the substance discovered that the effect of the decaying particles – the release of energy – could appear at the same moment at a distance from the site of the reaction, and that the location of the displaced reaction could be controlled. This was called the 'tele-explosion' or 'TX effect'. Exploiting this property, it was proposed that 'it would possible to produce a nuclear explosion that, transmitted with the speed of light, would release its destructive energies not where it was detonated, but at any location one chose on the globe' (Lem 1983: 139–140). It was the military implications of this discovery that led to the Pentagon taking over the Project, at the same time revealing the existence of their parallel project and confirming their distrust of the scientists' reliability with regard to their aim of security. By the time of the revelation and takeover, however, the scientists had also discovered the inbuilt safety mechanism, for the greater the energy released, the less accurate the localization of the effect, and so the effect could not be deployed as a weapon.

The biophilic effect and the TX-effect are of course fictions, serving the social satire. The important point is that, through them, Lem has identified a different kind of phenomenon: the productive or generative nature of the interaction between human endeavour and what, in the book, is classified in turn as real, or fiction, or error – the message. We have stepped beyond the alternatives of communication – potentially, their mind touching ours – or solipsism – the possibility that we are alone in the Universe. This principle of creativity (rather than representation) is however bought at the price of accepting the theosophical schema, introduced by Hogarth as he muses on the nature of the civilization that can emit signals promoting the development of life in the Universe whilst at the same time guarding against their misuse. This picture is elaborated in some details in the theories put forward by a cosmogonist and an astrobiologist – both imports from the parallel 'ghost' project – in the last chapter. The physicist speaks of the neutrino signal as the transmission of information between successive universes, separated by periodic catastrophes in which the laws of physics cease to operate, as the Cosmos expands and contracts. And the astrobiologist (surely an early use of the term) adds intentionality: an intelligent form of life, able to anticipate its end in the scheme of cosmic successions, had sought to influence the future state of the Universe and to control the cosmogonic process, making it favourable to the production of life (and so, ultimately, favourable to us).

Whether one accepts these constructions at scale, with their periodization of thirty billion years separating each successive universe, seems to be a matter of character. Hogarth suggests such matters are unknowable. He offers three grounds for his scepticism. First, the difference in intellectual and moral culture between a civilization that expends so much effort on promoting the good of others (assuming life to be a good) and ours, which devotes all its resources and intelligence to paranoia and defence, is too great to be crossed. Then, it is impossible, given the singularity of the evidence, to decide whether one is reading too much into that evidence, seeing intention where there is none, or, on the contrary, too little, missing patterns that are significant. And last, one may argue from our experience here on earth that it is impossible to gain an understanding of others, for all the indications we have suggest that what we take for human sympathy is the product of projecting our own preoccupations. In this fashion, Hogarth's misanthropy seals his solipsistic conclusion; he is not a man given to cooperative projects.

The contrast between the two accounts, which I termed at the outset optimistic and pessimistic, then turns out to be an opposition between two kinds of way of being in the world. One is confident and simplifying: theatrical, performative, and unself-conscious. The other complexifies things: it is novelesque, considering and reflective, concerned with absence, delay, and silence. They contain quite different attitudes to time; the one consists in a predictable sequence unfolding in linear time, the other in hesitations, repetitions, and returns to the past, making new connections and drawing new conclusions. This not simply to say that Lem is a better writer than Gunn, though his work is of quite a different order; it is to suggest that each captures a different aspect of the materials they are describing and exploring. There are different dimensions present in the space programme.

II. Later versions

The third novel, Carl Sagan's *Contact*, published in 1985, builds on Gunn but goes further, making new connections. The book sold well and was the basis of a successful film of the same name which came out in 1997. An astronomer, Sagan participated both in NASA and in the SETI programme from the earliest days; he was a participant in the Green Bank, West Virginia, workshop in 1961. He also played an important role as a public advocate for the exploration of space and as a writer and broadcaster advancing the public understanding of science. He is a valuable witness.

Sagan had laid out the accepted terms of the project of interstellar communication twenty years before publishing *Contact*. In Shklovsky and Sagan (1966), exploring components of the Drake equation, he suggested that 'the number of extant civilizations substantially in advance of our own in the Galaxy today appears to be perhaps between fifty thousand

and one million. The average distance between technical civilizations is between a few hundred light years and about one thousand light years. The average age of communicating technical civilizations is ten thousand years'. Now that we on Earth have made our presence known through the emission of radio signals, and 'if interstellar spaceflight by advanced technical civilizations is commonplace, we may expect an emissary, perhaps in the next several hundred years'. And he held out the prospect of our gaining 'full membership in a galactic community of societies'.[4]

In an earlier article, published in *Planetary and Space Science* (Sagan 1963), he elaborated the nature of this community, evoking the possibility of intergalactic exploration and an accompanying trade in goods and concepts, enhancing 'the vitality of the participating societies'. He also commented on the stability of this, for us, future commonwealth: 'It must be assumed that a highly advanced society would also be stable over very long periods, preserving the records of previous expeditions ... According to this hypothesis, civilizations throughout the galaxy probably pool their results ... There may be a central galactic information repository where knowledge is assembled, making it easier for those with access to such information to guess where, in the galaxy, newly intelligent life is about to appear ...' All this, we may remark, is pure Theosophy: Akashic records, centralized learning, a race of Masters helping organize and develop the Universe, the care for new intelligent species. This may have been learnt indirectly, from science fiction novels, but the parallels are striking. Given this perspective, it follows there may have been a sequence of extra-terrestrial visits to Earth in the past, perhaps '10,000 times over the full span of the earth's history', and evidence for some of these may appear in ancient texts (again, an argument proposed by theosophists).[5]

We may then be prepared for the direction Sagan develops Gunn's basic plot.[6] In the first half, he follows Gunn's narrative, though giving more detail at every point. The central figure, Ellie Arroway, an astronomer, is in charge

---

[4] These quotations are taken from Gunn (1972: 51, 217–218).
[5] See quotations in Gunn (1972: 268–269, 265, 270).
[6] Gomel discusses Sagan's approach in contrast to Lem's (see Gomel 2014: 13–18, 202–206).

of a listening array of radio telescopes, Project Argus, situated at Socorro in New Mexico. She is confronted with the problems of sustaining her team's confidence in the face of lack of positive evidence and of defending the project's funding from competition by other scientists for other, more glamorous, projects. An unambiguous signal is, however, received, with prime numbers acting as a beacon and, at a second level, the retransmission of an extremely early television broadcast. Immediately, the situation becomes more complex, with security and political interests coming into the game, together with collaboration with a wider group of scientists and increasing public interest and unease. Because both the science and the politics have an inescapable global dimension, an international consortium is set up. Rather than there being an exchange of messages between the sender and recipients, the Message (as it becomes known) proves to be self-sufficient and, eventually, self-interpreting, with the detection of a third and then a fourth level of coding. These are, respectively, a vast body of text, transmitted sequentially and repeatedly, and, eventually, the primer detected, needed to decipher the text unambiguously, which turns out to be scattered throughout the text.

The team of scientists realize they have been given complex technical instructions for the manufacture of a machine: blueprints and so forth, together with entire processes for making new equipment and component parts. National consortia are set up to build versions of what comes to be called the Machine, with new industries and technologies being developed from the instructions to achieve this. Trust and cooperation develop between the nations in pursuit of this project, not to mention the consumption of a vast part of national budgets previously given to nuclear arsenals. Despite this spirit of human brotherhood, there is a good deal of religious and other opposition to the project in the United States. These objections are not so much answered, as they are in Gunn, as circumvented; after a sabotage attack on the American project, the final stages of construction are moved to Japan and funded by private enterprise.

From this point on, the narrative changes direction. So far, compared with Gunn, Sagan's account benefits from fifteen years' more experience in SETI and the space programme, Sagan's insider knowledge of the organizational politics, and more scope to develop the story, including extensive instruction to the reader. Nevertheless, most real-world complexities are as absent as they were in the earlier version; Gunn's basic framework is continued.

Once the Machine starts to be constructed, however, we move into another register. In a word, the scientists and technologists working on the project have to take their instructions on faith. We are given an external description of the prominent features of the Machine (see Sagan 1985: chapter 15) but the function of each component is a mystery.

And so is the operation of the Machine. Five representative human beings take their places, including Ellie Arroway, and the apparatus is launched. Effectively, they are in a time machine, though we are given an account of a galactic transport system using the properties of black holes. After a series of transfers, likened to a metro system, they arrive at an enormous artificial satellite station and dock. Once disembarked, they find themselves in a simulated earth-like setting and sleep overnight. The next day, each separately encounters a representative of the civilization that has summoned them; let us call these the 'hosts'. The host meeting Ellie takes on the appearance of her beloved and long dead father. They converse, and the host reveals something of the scope of his advanced civilization and their purpose in making contact.

The hosts have entered the heads of the visiting humans during the previous night, while they slept, and have taken copies of their minds. Mind reading and telepathy, we understand, are achieved sciences. The host, then, need not ask Ellie concerning her thoughts and understanding, and this allows him to offer a discourse that anticipates her objections and answers her unspoken questions. He explains first the purpose of the mission to which he belongs,[7] 'We collect information', and offers an evaluation of the merits of the Earth's civilization: 'I think it amazing that you have done as well as you have. You've got hardly any theory of social organization, astonishingly backward economic systems, no grasp of the machinery of historical prediction, and very little knowledge about yourselves. Considering how fast your world is changing, it's amazing you haven't blown yourselves to bits by now. That's why we don't want to write you off just yet …'.

---

7   All quotations are to be found in Sagan (1985: chapter 20).

This last sentence hints at the possible purpose of the intervention: they (the hosts) can offer a little help to Earth's civilization, although 'there are certain limitations imposed by causality'. He offers her sight of their resources by a means that resembles astral travel, showing her the transport system between star systems, mapping out the metro system across the Galaxy, allowing her to locate the situation of their meeting, near the centre. And he explains the transport system's function; it is a distribution system, based on energy flows, and is, moreover, a 'cooperative project of many galaxies'. For the most part, the participating civilizations are engineers; only a few, as in the present case, are 'involved with emerging civilizations'. It becomes clear that the universe is cultivated and actively made. As Ellie remarks, inwardly, 'this was a hierarchy of beings on a scale she had not imagined'. We gain some insight into the earlier history of the energy transport system; the present participants – many species from many worlds – found these transit systems, left behind by a previous maker, a 'Galaxy-wide civilization' which has left no other traces. 'We're just caretakers', her host tells her, adding 'Maybe some day they'll come back'. We are introduced to the notion of a yet-higher kind, mysteriously absent, just as the hosts had been there but absent from our lives before contact was made.

We reach the final part of the purpose of the encounter, the mission to be imparted to the human representatives. There is a message built into the fabric of the universe which the Caretakers need help in deciphering, allowing them to grasp the underlying purpose of the whole. One clue to this message is to be found in analysis of the value of the mathematical constant, 'Pi', which turns out to be the product of eleven prime numbers, allowing transmission of a message in eleven dimensions, addressed to us.

In the final chapter, Ellie sums up what she has learnt. She reflects on 'the Caretakers with their network of tunnels through the Galaxy. They had witnessed and perhaps influenced the origin and development of life on millions of worlds. They were building galaxies, closing off sectors of the universe. They could manage at least a limited kind of time travel. They were gods ... But even they had their limitations. They had not built the tunnels and were unable to do so. They had not inserted the message into the transcendental number, and could not even read it. The Tunnel builders and the Pi inscribers were somebody else. They had left no forwarding address ...'.

Ellie rejoins the others, each accompanied by their own host. The five are returned to their craft, united in their insight that 'here were beings who live in the sky, beings enormously knowledgeable and powerful, beings concerned for our survival, beings with a set of expectations about how we should behave'. The craft returns them to their starting point, the launch pad on Earth.

Here, their problems accrue. The timing does not work; the journey was expected to take decades, but they find that only twenty minutes has elapsed in earth time. Their recording devices are all blank and they have brought back no incontrovertible evidence, only signs in the machine of 'tensile and compressional stresses'. These problems match those of flying saucer reports and close encounters. Worst of all, their accounts are not counted credible. Yet, although not remarked in the novel, this narrative is deeply familiar. We know about galactic Masters who display advanced technical, mental and moral knowledge, who oversee the development of the Universe and guide humanity's faltering steps, who help us overcome the crises we face because of our uneven progress, but at the same time confirm our central importance to the whole cosmic process: it is the theosophical account without remainder.

Sagan is completely conversant with the rules of the game. We will be able to encounter the dead and talk to them, and the facsimile will be indistinguishable from the original. In the last chapters, the narrator examines the question of eternal life, conceived as survival of the individual mind. More, Ellie ponders the identity of the organizing Cosmic mind, which can be revealed by decoding a mathematical constant, and the loving purposes of life intuited by human behaviour. And she reflects on the inevitability with which these insights will gain sway in a future human civilization, because God is a mathematician and, through the mathematics inscribed in the universe, humankind will come to share an unambiguous reading of the truth, on the basis of evidence that will be available to all. Scientific truth in the form of reliable prediction, a repeated trope, will replace faith because truth is an objective fact, but nothing of value contained in present faiths will be lost. We can see why the religious questions were not resolved in the earlier chapters, but just left to lie. Even though we are not sufficiently advanced to be central in this regard to the workings of the Cosmos, it seems we are the point at which the message is coming to be read by intelligent beings; we are – a familiar line – the coming to self-awareness of the entire cosmic process: 'Whoever makes the universe hides messages in transcendental numbers so they'll be read fifteen billion years later when intelligent life finally evolves'. That is the human calling.

This is a novel, and one should not read too much into it. It is nevertheless striking the ease with which one can move from an optimistic account of the possibility of communication to an apparatus of talking with the dead, communicating with ascended Masters who are attentive to our limitations and needs, and taking part in a vital crisis in the development of the Cosmos, its coming to self-awareness. We might contrast Lem's account of absence, delay, and silence with Sagan's recourse, in the context of the fading Cold War, to tropes drawn from American metaphysical religion, and we might recall the earlier suggestion that such a metaphysical hope of communication and the promise of a transparent scientific language may share a common historical root.

We now turn back to a recent instance of the solipsistic position.

Story of your life

A final case, which runs over some of the same issues with considerable originality, is Ted Chiang's novella 'Story of your life', first published in 1998.[8] It has served as the basis of the film *Arrival* (2016). If Lem's account gives the point of view of the natural scientist, Chiang gives the perspective of a scholar from the humanities. The story juxtaposes two themes, the involvement of a linguist, Louise Banks, in a government-sponsored negotiation with alien visitors, and a letter which Louise[9] writes to her daughter on the occasion of her birth. In the letter, the mother draws together episodes from her daughter's relatively short life – she dies in an accident at the age of 25 – because Louise is capable of seeing the whole span synoptically. One theme, then, concerns a sequence of events – from

---

8   There is a discussion of the story in Vint (2014: 136–140).
9   I follow the author's practice in using her first name.

the arrival of spaceships orbiting the Earth to their sudden departure – and the other an act of clairvoyance in which the narrator can recall every detail of her life considered as a whole, with every thought, word, feeling, even sensation, in its original state. The story is constructed around this contrast between linear time, a succession of events, and time experienced as a simultaneity, as a form or Gestalt with all the elements co-present. The two views of time are linked because Louise learns to experience her life as a whole as she becomes competent in the visitors' language and gains access to their worldview.

Since it is a short story, several elements are only sketched in; Louise is paired with a physicist, Gary, who permits certain explanations and analogies to appear when required, and the military sponsorship and international implications are to a large extent taken for granted. The main focus is on the language, or languages, learnt. A number of features appear which recall earlier discussions or develop a new aspect (in the case of writing).

First, Chiang resolves the problem of communication at a distance, on which other attempts have foundered, by allowing face-to-face communication. The alien spaceships appear in orbit and scatter a series of objects across the world: 'artefacts appeared in meadows'. These artefacts are two-way screens which allow communication in real time, using both sight and sound, between Earth and the spaceships. This allows description of the aliens, which display a seven-fold symmetry, with seven limbs and seven eyes, and are termed heptapods by their investigators. Louise and Gary are selected by the Army to establish contact with a pair of heptapod interlocutors; there are similar interviews going on in other places.

This face-to-face situation allows Louise to begin as she would when dealing with a human group with another language (although she regrets the absence of a bilingual translator); she begins by showing and telling, indicating an object and then giving a sound, in this fashion establishing identity between sound and thing. The heptapods cooperate and reciprocate, mirroring her procedures. Together, they begin with nouns – naming species and body parts, giving personal names – and then move to verbs, first intransitive actions – jumping, speaking, walking – and then transitive – changing the object eaten, for example. This work proceeds exactly as if the heptapods were another human tribe, establishing basic phonetics, vocabulary, and syntax. Only after establishing the basis of communication can they turn to more complex matters, comparing bodies of knowledge and approaches to physics. So far, we might say, this is reminiscent of Freudenthal, who has to use ideograms to substitute for showing and telling in order to start the learning process and is apparently untroubled by the possibility of indeterminacy of reference (how can I be sure you mean the same as I do when we both say X?).

Second, since she is dealing with an advanced technological civilization, Louise quickly turns to writing, feeling it may be easier to identify significant units – graphemes rather than phonemes – in this way. Again, the heptapods cooperate; each team sets up the equivalent of a white board and run their verbal enquiries in parallel with inscription, producing two bodies of material. Louise works out that, while the spoken language is sequential, the writing is not; it presents no order, for each inscription can only be read as a whole, marking relationships but without temporal priority; they have no marking for cause and effect, for example. She terms this a 'semasiographic writing system' – signs that convey meaning without reference to speech – which has its own rules for constructing sentences.[10] The heptapod writing system is a 'full-fledged, general purpose graphical language', offering a 'grammar in two dimensions', similar in this regard to the systems of notation used for mathematical equations or music or dance, but with far greater sophistication and flexibility than any human products. Louise calls this writing system Heptapod B, in distinction to the spoken language, Heptapod A. It emerges that the writing is the clue to the worldview of the heptapods, and that its peculiarities determine many of the properties and puzzles of the spoken language, with its free ordering of words and even clauses and the use of 'many levels of center-embedding of clauses'.[11] Writing has priority over speech.

---

[10] International signs, such as the conventional 'stop' sign, are examples. This kind of sign system becomes important when considering, for instance, the burial of atomic waste and putting up warning signs that will be understood in the far future; see the discussion in Oberhaus (2019: 137–139).

[11] Humans cannot readily invert the order of clauses in conditional sentences, for example, and even one embedded clause in English – 'the man the boy spoke to is a

Third, a clue to the worldview is given by the discussions of physics, which are central because the military and state sponsors of the encounter hope to gain new scientific and technical know-how. The breakthrough in mutual understanding comes not in comparing approaches to geometry or algebra – both straightforward branches of mathematics in human terms – but when considering an account cast in terms of maxima and minima, using the case of Fermat's principle of least time (the proposal that light takes the fastest possible route between two points). It appears heptapod mathematics is based on 'variational principles'. Louise subsequently explores the concept: a variational principle is not causal in form but 'purposive, almost teleological', as if the light beam follows a command to minimize the time taken to reach a destination. On this model, 'the ray of light has to know where it will ultimately end up before it can choose the direction to begin to move in'. In parallel fashion, she realizes, 'the heptapod had to know how the entire sentence would be laid out before it could write the very first stroke'. In short, while we tend to think in parts, in terms of cause and effect (if x, then y), they start with the whole; we think in sequences, they think in simultaneities.

We might notice in passing that the two mathematical systems – human and heptapod – are taken to be ways of describing an independently existing physical universe. And that humans at least can recast their mathematics in terms of the other point of view, even if not by intuition, for almost all physical laws can be restated as variational principles; action, for example, can be seen as the difference between kinetic and potential energy integrated over time. What they see by intuition, we use calculus to grasp. A problem (not to be followed up) concerns the nature of the wholes with which the heptapods are at ease: how are they defined? Where are they found? Is it practical to reverse what for us is a retrospective framing and to imagine that the frame comes first? This priority supports the mathematical Platonism adopted whereby perception plays no role in the creation of the objects apprehended.

---

friend of mine' – is relatively rare; cases of multiple embedding are said not to work because of the frailty of human short-term memory.

Fourth and last, unlike the majority of authors with which we are concerned, Chiang has recourse to a version of the Sapir-Whorf theory (recalling Delaney and aspects of Lem). This recourse can be distinguished from his idea of encounter being situated in specific situations and engagements, rather than the communication of ideas at a distance, although the practical embedding lends support to the adoption of the theory. So, as Louise learns to write Heptapod B, she finds it is 'changing the way … [she] thought'. This alteration has a number of aspects. In the first place, her thoughts become graphically coded so that, instead of expressing them to herself in an inner voice, she sees the written form in her mind's eye. Instead of meditating action, she contemplates symmetries, with premisses and conclusions interchangeable.

This practice, next, involves a different perspective. Humans intuitively see the world in terms of cause and effect, and so as a succession of events. Heptapods, however, find the world meaningful as a pattern over time, interpreting the world in terms of maxima and minima, seeing the purpose in a stream of events. The heptapod perspective demands knowledge of both the initial and final states in order to understand; for this reason, they observe and offer no initiative. In practice, they only feed back to their human interlocutors what they have learned from them.

Thus far, one might suppose that Louise is talking about perceiving patterns, or the 'penny dropping'. But she now turns to examine some more unusual properties of mind associated with the perspective, describing how, as a learner, one comes to see the whole so that one has knowledge of one's entire life. This is the property that allows the accompanying second theme, the letter to her daughter which treats of their common life as a single, known episode – a totality. Having learned to think in Heptapod B, Louise gains access to all her 'memories', from the period of the interviews with the heptapod informants to the time of her death. Her 'memories grew like a column of cigarette ash, laid down by the infinitesimal sliver of combustion that was my consciousness, marking the sequential present'. The memories fell into place in blocks and became composed as a whole. Moreover, these memories from life are present in every detail, with every sensation recalled as it was experienced, comparable to the way an infant lives in the instant. Nothing is lost; Louise is clairvoyant and utterly alive, as her past and future live in her present.

Further to that ability to call up or recall, Louise also speaks of not simply living with knowledge of her future and her past, but sometimes of having glimpses of the whole future and past altogether: 'my consciousness becomes a half-century-long ember burning outside time'. She experiences the entire epoch of her life as a simultaneity. This is a Huxley-like state of transcendence, the simultaneous consciousness of everything within the compass of a human life without loss (cf. Aldous Huxley's *The Perennial Philosophy*, 1946). It might with practice be possible to go further in heptapod wisdom and to join in the Cosmic Consciousness.

Louise also explores the contrast between their experience and ours in more abstract, philosophical terms around the question of free will and determinism. If you know the future, by clairvoyance and time travel (although Chiang avoids these terms), you might in principle frustrate it. She resolves this problem by appealing to the completeness of each perspective (the strong version of the Sapir-Whorf hypothesis): 'every physical event was an utterance that could be parsed in two entirely different ways, one causal and the other teleological, both valid, neither one disqualifiable no matter how much context was available'. In one condition, one has no thought of the other state; in the one state, one seeks to effect changes, in the other, one embraces the processes discerned. In brief, the heptapods seek to 'enact chronology' and, if you know the future, you neither act contrary to it nor do you talk about your knowledge.

Louise elaborates her understanding of features of the heptapod worldview. For the heptapods, language is not descriptive nor is it used to convey information, for they already know what would be said. 'Instead of using language to inform, they use language to actualize'; in their world, 'for their knowledge to be true, the conversation would have to take place'. She describes this as a 'performative' view of language, although she sees this in terms of giving a performance: speaking is performing. For this reason, ritual is an important part of their life, which is also sustained by giving gifts rather than economic exchange. More, the gifts turn out to be variations of gifts given to them; the physics they impart is never new, although the State Department people hope for a miracle such as a space drive or the secret of cold fusion. And, after a period, like the Trobriand Islanders they resemble, they depart without explanation; their screens go blank and their ships leave Earth's orbit. We might wonder whether they have moved on to another island in their cycle of reciprocal gift giving.

Chiang's story differs in several regards from the earlier novels. Its setting is post-SETI and post-Cold War and bears no trace of the 'war on terrorism'. It is contemporary to the development of astrobiology but makes no reference to it. Some of the factors central to SETI, such as the problems of communication at vast distance, are side stepped, and others, such as the basis allowing intercommunication, are taken for granted: we are given aliens near at hand (but behind screens), interested in humans and capable of communication. In a similar fashion, we assume the perfection of the communications technology provided; the presence of state and military interests and their managerial role are taken as given but have no significance; and the centrality of mathematics and physics to the exchanges is a constant. The frame has changed, and the old elements given new values; yet the two limits, the possibility of solipsism and the hope of direct communication, remain.

The plot overall resembles Heinlein's story and is a comparable achievement in smaller scope: an exploration of human potential dressed up by introducing some itinerant extra-terrestrials passing mental goods between parties. The underlying inspiration has changed; the focus on writing systems is contemporary, as is the linking of the opposition between writing and speech to a series of other contrasts: teleological/causal, synchronic/diachronic (simultaneity/sequence), whole/parts. But the underlying message is much the same. It explores contemporary shifts in explanatory schemata, using mathematics and physics to mark the changes, which allow recourse to notions of telepathy, recovered memory, and cosmic consciousness as credible ways of understanding aspects of human experience under these new conditions.

In this account, despite its commitment to the power of language to shape thought and action, we imagine a universe made up of independent, objective events that are open to either causal or teleological interpretation. This is despite the fact we start from the necessity of co-presence and interaction, or local particularity. For Louise, the two worldviews – one characterized by the sequence of events, the other by a simultaneity of elements – are distributed one to each party. We have split the resources of the twentieth-century western mind in two, with positivism on the one hand and a version of the Cosmic Mind, the simultaneous consciousness of everything, on the other. In this split, we lose sight of both the autonomy of the 'event', when the mode of registration alters along with the 'thing' on the ground, and human responses to this double shift. For Chiang, the heptapods remain inscrutable, and significant change takes place only within the human sphere. Once again, we glimpse the creativity of human intelligence which Lem parodies in the biophilic and the T-X effects. In both instances, however, although there are traces of a third position, we are controlled in the main by the dream of direct communication and the realization of its impossibility.

III. A balance sheet

We began with the coincidence in time of the Air Force shutting down its unit investigating flying saucer reports and NASA taking seriously the prospect of searching for extra-terrestrial intelligence, reading both in terms of the realization of forms created initially in Air Force circles by the beginning of the 1950s. The two contemporary stories provide a contrast, although, in practice, each follows the same narrative arc of initial interest followed by accumulating impediments. It is clear that these were independent operations, each with their own rationale, but that underlying and shaping each was an engagement with a certain conception of communication, conceived as conveying information, or failing to do so, so that the first decision can be seen in the perspective of the limitations, and the second of the hopes, of that conception. Investigation of this concept reveals at its heart a double ambition, first, of creating an objective scientific language capable of transmitting unambiguous information between parties and, second, of moral employments of that power of unequivocal transmission to beneficial human ends. The history of communication in its modern sense lies in an interplay between these ambitions, as the potential of each new technology is mapped in terms drawn from a metaphysical vocabulary, and each new human situation understood in concepts drawn from the technological field.

Taking a longer historical perspective, this ideal of communication as the distinction of 'message' from 'noise' can be seen to tie up both with the history of certain scientific ideas – in particular, action at a distance – and with attempts to make such discoveries relevant to common life through Mesmerism and, above all, Spiritualism, bringing us in turn back to theosophical concerns. Indeed, the complex of ideas around the concept of communication that concerns us took form in precisely the period and circumstances to which Theosophy provided some kind of ethical summary and description. We can trace the characteristic concerns of this complex in terms of the hope of direct communication (mind to mind) and the fear of its impossibility, and identify two periods in the twentieth century, in the aftermath of each World War, when such concerns appeared acutely felt and were elaborated.

Reviewing explorations of the characteristics and resources of alien languages in science fiction and in projects to construct a language for communication with other civilizations in the context of SETI, together with science fiction novels that focus on 'contact' in that context, confirms both the periodization and the model of communication with its resources and limitations. While Heinlein's *Stranger* allows a description of the earlier form, we have focussed for the most part on the later, post-War productions, which remain within the confines of the model identified.

Both the earlier and the later science fiction stories hover selfconsciously between two worlds. On the one hand, they explore the possibilities of a 'discorporate' world (to borrow Heinlein's term), without hypocrisy, jealousy, conflict, or war, where minds communicate without impediment and act directly on one another. At the same time, they play with ideas of influence and the interaction of minds with matter, and of the mixing of bodies in both love and hostile contact. They extend the discussion in a significant way. On the other hand, they continue to track the possibility that nothing will emerge from all this labour; that, from distance, different degrees of incommensurability, or simple absence, no contact can be made.

The discussion of 'real' artificial languages draws on many of the same concerns. Here too we meet with a human-centred view of the universe, for there are traces of a teleological perspective in the idea of contact with advanced technical civilizations, desiring communion with us and tending to mental rather than bodily communication. There is an aspect of time travelling in this notion: of meeting forms of our future selves. Similarly, there is a notion of harmony beneath the surface in the matching of the structure of the nervous system (the brain) with the computer, and in the privileging of reason, in the dream of a universal language which of necessity reduces language to information transmitted and stored. There is perhaps less self-consciousness in the scientific models than in the science fiction explorations. And the same fears persist, perhaps more acutely.

Can we go further? For the model leaves open the question of whether there might be other ways of mapping the territory. From the discussion of the 'contact' novels in particular, we can see that new powers of human creativity tend to be associated with the appearance, or even just the near approach, of intelligent aliens. While supernormal powers are attributed to alien minds, their practical effects are to draw out new discoveries and perspectives from the humans involved. These possibilities are emphasized particularly by Lem and Chiang. We might say that aliens act as relays, joining different times and distant places, allowing new properties to emerge in hitherto confined zones of human activity. These practical effects are expressed in both positive and negative fashion, allowing new relationships and possibilities to form, but also cutting off older ties.

In short, the two linked histories we identified – Air Force, SETI – allow us to pursue the continuing life of the categories invented earlier, in the immediate post-War period. These histories point us to longer-term considerations which reveal the powers and limits of the model being exposed, and these can be investigated in detail through a series of writings, both fiction and non-fiction.

And these representations also appear in civilian reports of flying saucers, beyond the direct influence of the military, the space industry, and their science fiction epigones, offering renewed life to these forms in new settings, expressing in multiple ways the conventions and energies that appear in these Martian languages.

Bibliography

Albanese, Catherine, *A Republic of Mind & Spirit: A Cultural History of American Metaphysical Religion*, New Haven, Yale University Press, 2007.
Balzac, Honoré de, *Ursule Mirouët*, London, Dent and Sons Ltd., 1947 [1841].
Billingham, John, 'SETI: The NASA Years', in Douglas Vakoch (ed.), *Archaeology, Anthropology, and Interstellar Communication*, Washington DC, NASA Office of Communication (The NASA History Series), 2014: 38–69.
Billings, Linda, 'Astrobiology in Culture: The Search for Extraterrestrial Life as "Science"', *Astrobiology* 12 (10), 2012: 1–12.
Bilstein, Roger, *Orders of Magnitude: A History of the NACA and NASA, 1915–1990*, Washington, DC, NASA Scientific and Technical Information Division, 1989.
Bowler, Peter, *The Eclipse of Darwinism: Anti-Darwinian Evolution Theories in the Decades around 1900*, Baltimore, Johns Hopkins University Press, 1992 [1983].
Cantril, Hadley, *The Invasion from Mars: A Study in the Psychology of Panic*, New Brunswick, Transaction Publishers, 2008 [1940].
Carey, John, *The Intellectuals and the Masses: Pride and Prejudice among the Literary Intelligentsia, 1880–1939*, London, Faber & Faber, 1992.
Chase, Stuart, *The Tyranny of Words*, New York, Harcourt Brace Jovanovich, 1966 [1938].
Chiang, Ted, 'Story of Your Life', in Ted Chiang, *Stories of Your Life and Others*, New York, Vintage Books, 2002: 91–146.
Clark, Jerome, *The UFO Book: Encyclopedia of the Extraterrestrial*, Detroit, Visible Ink Press, 1998.
Conway Morris, Simon, *From Extraterrestrials to Animal Minds: Six Myths of Evolution*, West Conshohocken, PA, Templeton Press, 2022.
Cusack, Carole, *Invented Religions: Imagination, Fiction and Faith*, Farnham, Ashgate, 2010.
Davies, Paul, *The Eerie Silence: Are We Alone in the Universe?* London, Allen Lane, 2010.
Dean, Jodi, *Aliens in America: Conspiracy Cultures from Outerspace to Cyberspace*, Ithaca, NY, Cornell University Press, 1998.
Delaney, Samuel R., *Babel-17*, London, Gollancz, 2010 [1967].
Depew, David and Bruce Weber, *Darwinism Evolving: Systems Dynamics and the Genealogy of Natural Selection*, Cambridge, MA, MIT University Press, 1995.

Descombes, Vincent, *The Mind's Provisions: A Critique of Cognitivism*, Princeton, Princeton University Press, 2001 [1995].
Devereux, George, *Psychoanalysis and the Occult*, London, Souvenir Press, 1974 [1953].
DeVito, Carl L., *Science, SETI and Mathematics*, New York, Berghahn, 2014.
DeVito, Carl L. and Richard T. Oehrle, 'A Language Based on the Fundamental Facts of Science', *Journal of the British Interplanetary Society* 43, 1990: 561–568.
Dick, Steven J., *Life on Other Worlds: The 20th-Century Extraterrestrial Life Debate*, Cambridge, Cambridge University Press, 1998.
Dick, Steven J., 'The Search for Extraterrestrial Intelligence and the High-Resolution Microwave Survey', *Space Science Reviews* 64, 1999: 93–139.
Domagal-Goldman, Shawn and Katherine Wright (eds), 'The Astrobiology Primer v2.0', *Astrobiology* 16 (8), 2016: 561–653.
Doyle, Richard, 'Close Encounters of the Nth Kind: Becoming Sampled and the Mullis-ship Connection', in Debbora Battaglia (ed.), *E.T. Culture: Anthropology in Outerspaces*, Durham, Duke University Press, 2005: 200–217.
Eghigian, Greg, 'Making UFOs Make Sense: Ufology, Science, and the History of Their Mutual Mistrust', *Public Understanding of Science*, 26 (5), 2015: 1–15.
Ellenberger, Henri, *The Discovery of the Unconscious: The History and Evolution of Dynamic Psychiatry*, New York, Basic Books, 1970.
Fawcett, Lawrence and Barry Greenwood, *Clear Intent*, Englewood Cliffs, NJ, Prentice-Hall, 1984.
Franklin, H. Bruce, *Robert A. Heinlein: America as Science Fiction*, New York, Oxford University Press, 1980.
Freud, Sigmund, 'Civilization and Its Discontents [1929]', in Sigmund Freud, *Civilization, Society and Religion. The Penguin Freud Library*, vol. 12, Harmondsworth, Penguin, 1991: 251–340.
Freudenthal, Hans, 'Excerpts from *Lincos: Design for a Language for Cosmic Intercourse*', in Edward Regis (ed.), *Extraterrestrials: Science and Alien Intelligence*, Cambridge, Cambridge University Press, 1985: 215–228.
Freudenthal, Hans, *Lincos: Design for a Language for Cosmic Intercourse*, Amsterdam, North-Holland Publishing, 1960.
Fuller, Robert, *Mesmerism and the American Cure of Souls*, Philadelphia, University of Pennsylvania Press, 1982.
Garber, Stephen, 'A Political History of NASA's SETI Program', in Douglas Vakoch (ed.), *Archaeology, Anthropology, and Interstellar Communication*, Washington DC, NASA Office of Communication (The NASA History Series), 2014: 70–106.

Garber, Stephen, 'Searching for Good Science: The Cancellation of NASA's SETI Program', *Journal of the British Interplanetary Society* 52, 1999: 3–12.
Gomel, Elana, *Science Fiction, Alien Encounters, and the Ethics of Posthumanism: Beyond the Golden Rule*, Basingstoke, Palgrave Macmillan, 2014.
Grinspoon, David, *Lonely Planets: The Natural Philosophy of Alien Life*, New York, Harper Collins, 2004.
Gunn, James, *The Listeners*, Tiburon, CA, Reputation Books, 2017 [1972].
Hall, Richard, *The UFO Evidence*, vol. 2, Lanhan, MD, Scarecrow Press, 2000.
Hanson, Miriam Bratu, 'Introduction', in Siegfried Kracauer, *Theory of Film: The Redemption of Physical Reality*, Princeton, NJ,, Princeton University Press, 1997.
Hayles, N. Katherine, *How We Became Posthuman: Virtual Bodies in Cybernetics, Literature, and Informatics*, Chicago, University of Chicago Press, 1999.
Heidmann, Jean, *Extraterrestrial Intelligence*, Cambridge, Cambridge University Press, 1995 [1992].
Heinlein, Robert, *Stranger in a Strange Land*, London, Hodder and Stoughton, 2005 [1960].
Hesse, Mary, *Forces and Fields: The Concept of Action at a Distance in the History of Physics*, New York, Dover, 2005 [1961].
Hoggart, Richard, *The Uses of Literacy*, London, Chatto and Windus, 1957.
Hoyt, Diana, *UFOCRITIQUE – UFOs, Social Intelligence and the Condon Committee*, thesis, Virginia Polytechnic Institute and State University, 2000.
Husserl, Edmund, 'Philosophy and the Crisis of European Man' [1936], in Edmund Husserl, *Phenomenology and the Crisis of Philosophy*, Quentin Lauer (ed.), New York, Harper Torchbooks, 1965: 149–192.
Hutchins, Edwin, *Cognition in the Wild*, Cambridge, MA, The MIT Press, 1995.
Huxley, Aldous, *Brave New World Revisited*, London, Chatto and Windus, 1958.
Huxley, Aldous, *The Perennial Philosophy*, London, Chatto & Windus, 1946.
Huxley, Julian, *Evolution: The Modern Synthesis*, London, George Allen & Unwin Ltd., 1942.
Hynek, J. Allen, *The UFO Experience: A Scientific Enquiry*, New York, Ballantine Books, 1972.
Jacobs, David Michael, *The UFO Controversy in America*, Bloomington, Indiana University Press, 1975.
James, William, *The Varieties of Religious Experience: A Study in Human Nature*, London, Longmans, Green and Co., 1944 [1902].
Jameson, Fredric, *Archaeologies of the Future: The Desire Called Utopia and Other Science Fictions*, London, Verso, 2005.
Jenkins, Timothy, *Of Flying Saucers and Social Scientists: A Re-reading of When Prophecy Fails and of Cognitive Dissonance*, New York, Palgrave Macmillan, 2013.

Keane, Webb, *Christian Moderns: Freedom and Fetish in the Mission Encounter*, Berkeley, University of California Press, 2007.
Keyhoe, Donald, *Flying Saucers from Outer Space*, New York, Henry Holt & Co., 1953.
Kittler, Friedrich, 'Media and Drugs in Pynchon's Second World War' (1987), in Friedrich Kittler, *The Truth of the Technological World*, Stanford, Stanford University Press, 2013: 84–98.
Korzybski, Alfred, *Science and Sanity*, Fort Worth, TX, Institute of General Semantics, 2005 [1933].
Kripal, Jeffrey, *Esalen: America and the Religion of No Religion*, Chicago, University of Chicago Press, 2008.
Lasch, Christopher, *The Culture of Narcissism: American Life in an Age of Diminishing Expectations*, New York, W. W. Norton 1979.
Latour, Bruno, *We Have Never Been Modern*, Cambridge, MA, Harvard University Press, 1993.
Leavis, Q. D., *Fiction and the Reading Public*, London, Chatto & Windus, 1968 [1932].
Lederberg, Joshua, 'Exobiology: Approaches to Life Beyond the Earth', *Science* 132, 1960: 393–400.
Leinster, Murray, 'First Contact', in Stanley Schmidt (ed.), *Aliens from Analog*, New York, The Dial Press, 1983: 13–36 [1945].
Lem, Stanislaw, *Fiasco*, New York, Harcourt Brace Jovanovich, 1987 [1986].
Lem, Stanislaw, *His Master's Voice*, London, Secker & Warburg, 1983 [1968].
Lem, Stanislaw, *Microworlds: Writing on Science Fiction and Fantasy*, Franz Rottensteiner (ed.), San Diego, Harcourt Brace Jovanovich Publishers, 1984.
Lem, Stanislaw, *Solaris*, London, Faber & Faber, 1970 [1961].
Lethem, Jonathan, 'On Stanislaw Lem', *London Review of Books* 44 (3), 10 February 2022: 27–32.
Lovecraft, Howard Phillips, *Supernatural Horror in Literature*, New York, Dover 1973 [1945].
Luckhurst, Roger, *Science Fiction*, Cambridge, Polity Press, 2005.
Lyotard, Jean-François, *The Inhuman*, Stanford, Stanford University Press, 1991.
Mayr, Ernst, 'The Probability of Extraterrestrial Intelligent Life', in Edward Regis (ed.), *Extraterrestrials: Science and Alien Intelligence*, Cambridge, Cambridge University Press, 1985: 23–30.
McConnell, Brian, *Beyond Contact: A Guide to SETI and Communicating with Alien Civilizations*, Sebastopol, CA, O'Reilly and Associates, 2001.
McDougall, Walter, … *The Heavens and the Earth: A Political History of the Space Age*, Baltimore, The Johns Hopkins University Press, 1997 [1985].

Mills, C. Wright, *The Power Elite*, New York, Oxford University Press, 1956.
Minsky, Marvin, 'Communication with Alien Intelligence', *Byte Magazine*, April 1985. [= Minsky 1985a]
Minsky, Marvin, 'Why Intelligent Aliens Will Be Intelligible', in Edward Regis, Jr (ed.), *Extraterrestrials: Science and Alien Intelligence*, Cambridge, Cambridge University Press, 1985: 117–128. [= Minsky 1985b]
Morowitz, Harold and Eric Smith, *The Origin and Nature of Life on Earth: The Emergence of the Fourth Geosphere*, Cambridge, Cambridge University Press, 2016.
Nozick, Robert, 'R.S.V.P. – A Story', in Edward Regis, Jr (ed.), *Extraterrestrials: Science and Alien Intelligence*, Cambridge, Cambridge University Press, 1985: 267–273.
Oberhaus, Daniel, *Extraterrestrial Languages*, Cambridge, MA, MIT Press, 2019.
Ogden, C. K. and I. A. Richards, *The Meaning of Meaning: A Study of the Influence of Language upon Thought and of the Science of Symbolism*, New York, Harcourt, Brace and Co., 1923.
Ollongren, Alexander, *Astrolinguistics: Design of a Linguistic System for Interstellar Communication Based on Logic*, New York, Springer, 2013.
Orwell, George, *Nineteen Eighty-Four*, London, Secker and Warburg, 1949.
Owen, Alex, *The Darkened Room: Women, Power, and Spiritualism in Late Victorian England*, Chicago, University of Chicago Press, 2004 [1989].
Peebles, Curtis, *Watch the Skies! A Chronicle of the Flying Saucer Myth*, Washington, DC, Smithsonian Institution Press, 1994.
Peters, John Durham, *Speaking into the Air: A History of the Idea of Communication*, Chicago, University of Chicago Press, 1999.
Podmore, Frank, *Modern Spiritualism: A History and a Criticism*, 2 vols, New York, Cambridge University Pres, 2011 [1902].
Project Blue Book, *Special Report No. 14*, Air Technical Intelligence Center, Wright-Patterson AFB, OH, 1955.
Regis, Edward, Jr (ed.), *Extraterrestrials: Science and Alien Intelligence*, Cambridge, Cambridge University Press, 1985.
Rieff, Philip, *The Triumph of the Therapeutic: Uses of Faith after Freud*, Chicago, University of Chicago Press, 1966.
Riesman, David, *The Lonely Crowd, a Study of the Changing American Character*, New Haven, Yale University Press, 2001 [1950].
Roberts, Adam, *The History of Science Fiction*, London, Palgrave Macmillan, 2016.
Roth, Christopher, 'Ufology as Anthropology: Race, Extraterrestrials, and the Occult', in Debbora Battaglia (ed.), *E.T. Culture: Anthropology in Outerspaces*, Durham, Duke University Press, 2005: 38–93.

Russell, Mary Doria, *The Sparrow*, London, Black Swan, 1997 [1996].
Sagan, Carl, *Contact*, New York, Simon & Schuster, 1985.
Sagan, Carl, 'Direct Contact among Galactic Civilizations by Relativistic Interstellar Flight', *Planetary and Space Science* 11 (5), 1963: 485–498.
Sagan, Carl et al., *Murmurs of Earth: The Voyager Interstellar Record*, New York, Ballantine Books, 1978.
Samuels, David, 'Alien Tongues', in Debbora Battaglia (ed.), *E.T. Culture: Anthropology in Outerspaces*, Durham, NC, Duke University Press, 2005: 94–129.
Saunders, David and Roger Harkins, *UFOs? Yes!*, New York, World Publishing, 1969.
Schmitt, Carl, *Political Theology. Four Chapters on the Concept of Sovereignty* [1922, 1924], Cambridge, MA, MIT Press, 1988.
Scoles, Sarah, *Making Contact: Jill Tarter and the Search for Extraterrestrial Intelligence*, New York, Pegasus Books, 2017.
Sennett, Richard, *The Fall of Public Man*, New York, Alfred Knopf, 1977.
Shannon, Claude and Warren Weaver, *The Mathematical Theory of Communication*, Urbana, University of Illinois Press, 1948.
Shaver, Richard S., *I Remember Lemuria* (1945), reprint, LaVergne, Tennessee, 2016.
Shklovsky, Josif and Carl Sagan, *Intelligent Life in the Universe*, San Francisco, Holden-Day, 1966.
Stephenson, Neal, *Snow Crash*, London, Penguin Books, 2011 [1992].
Suvin, Darko, *Metamorphoses of Science Fiction: On the Poetics and History of a Literary Genre*, Oxford, Peter Lang, 2016 [1979].
Swords, Michael and Robert Powell (eds), *UFOs and Government: A Historical Inquiry*, San Antonio, Anomalist Books, 2012.
Vint, Sherryl, *Bodies of Tomorrow: Technology, Subjectivity, Science Fiction*, Toronto, University of Toronto Press, 2007.
Vint, Sherryl, *Science Fiction: A Guide for the Perplexed*, London, Bloomsbury, 2014.
Weber, Max, 'Science as a Vocation' [1918], in H. H. Gerth and C. Wright Mills (eds), *From Max Weber: Essays in Sociology*, London, Routledge, 1991: 129–156.
Wells, H. G., *The War of the Worlds*, Harmondsworth, Penguin Books, 1962 [1897].
Wilson, Edmund, *Classics and Commercials: A Literary Chronicle of the 1940s*, New York, Farrar, Straus and Giroux, 1950.

Index

action at a distance 49, 53, 57, 59, 142
Air Force, United States 6–17, 20, 141–143
alien life forms 56, 61, 73, 84–95, 111–115, 135, 143
animal magnetism 50, 55 *see also* Mesmerism
anti-humanism 69, 74, 89–96
APRO (Aerial Phenomena Research Organization) 15
ARPA (Advanced Research Projects Agency) 7, 24, 26
astrobiology 30, 34–43, 58
Astrolinguistics 104

Basic English 66, 77
Billingham, John 27–33
broadcast of *War of the Worlds* 61

Cantril, Hadley 61
CETI (Communication with Extra-Terrestrial Intelligence) 29–30
Chase, Stuart 77, 80
Chiang, Ted 134–141
Chomsky, Noam 84–88, 111 *see also* language, theories of
clairvoyance 50, 53, 84, 97, 135, 139
Clark, Jerome 13, 16
Close Encounters 8, 10, 133
Cocconi and Morrison 27–28, 63–64, 120
Cold War 3, 11, 17–19, 25, 33–34, 95, 134
Colorado Project 6, 12–17
communication 17, 23, 28, 45–72, 115–117

extra-terrestrial 25–27, 35, 37, 39, 97–109
    failures of 41–42, 45–70, 83
    history of modern form 48–50, 50–53, 65–70
    interstellar 30, 62, 99, 101, 103–104, 107–109, 114–115 119, 128 *see also* Cocconi and Morrison
    mind-to-mind 1, 47, 53–54, 59, 73, 83–84, 86, 97, 100, 106, 108–109, 120, 122, 142 *see also* telepathy
    and science fiction 72–76
    spiritualist inheritance 53–55
    two forms 45–48, 71–72
Condon, Edward 13–14, 16
Condon report 6, 12–17
Congressional interest in UFOs 9–10, 15
Conway Morris, Simon 64, 113
CUFOS (Center for UFO Studies) 15

Darwin, Charles 40
Dean, Jodi 23
Delaney, Samuel (*Babel-17*) 84–88, 138
Depew, David and Bruce Weber 36, 41
DeVito, Carl 104–106
Dick, Steven J. 32, 37
Drake equation 28, 36, 38, 128
Drake, Frank 27, 103, 120–121

ETH (Extra-Terrestrial Hypothesis) 16
exobiology 29–30, 42
exoplanets 37, 42 *see also* planetary detection
'extra-terrestrial' (appearance of the term) 15–16

'first contact' 57, 119–143
Freudenthal, Hans 99–102, 105–106, 110, 136

General Semantics 76–79
Gomel, Elana 90–95
gravitation 49–50
Green Bank (National Radio Astronomy Observatory) 28, 103, 128
Gunn, James 120–123

Heinlein, Robert 79–84
Hoyt, Diana 13, 16
Huxley, Aldous 69, 139

information theory 62–63, 65, 67–68, 70
interplanetary hypothesis (elimination of) 1, 5, 64
interplanetary signals 27–31, 63, 98–99, 102–107, 110, 115, 127, 129

Jacobs, David M. 6, 8–10, 13

Keyhoe, Donald 8–9
Kittler, Friedrich 17, 26, 71–72
Korean War 20–22
Korzybski, Alfred 76–79, 84, 102

language, theories of 84–88 *see also* Chomsky, Noam; Sapir-Whorf
languages, extra-terrestrial (fictional and non-fictional) 71–117
Lem, Stanislav 91–92, 123–128, 138, 141
liberal Protestant thought 80
Lincos 99–102, 104, 106–107
Locke, John 48–49, 54, 108, 114

Martian life and language 79–84
Martian linguistics, definition 1, 5, 59
Mayr, Ernst 38–39
McDougall, William 17–27, 29

Mesmer, Franz 49–50
Mesmerism 49–51, 53–54, 58, 142 *see also* animal magnetism
metaphysical religion 49, 51–53, 80, 134, 142
METI (Messaging Extra-Terrestrial Intelligence) 110, 116
mind
    cognitivist approaches 68, 78, 91, 111–114
    human and animal 92, 112–113
Minsky, Marvin 111–115
Modern Synthesis 40–41
MUFON (Mutual UFO Network) 15

NACA (National Advisory Committee for Aeronautics) 7, 19, 24
NAS (National Academy of Sciences) 28–29
NASA (National Aeronautics and Space Administration) 7, 17–33, 37, 42, 120
National Aeronautics and Space Act (1958) 23–24
National Research Council 30
negative theology 92–93
neurolinguistic programming 85–86
NICAP (National Investigations Committee on Aerial Phenomena) 8–11, 15–16
novels of first contact 119–143

Oberhaus, Daniel 98–117
obstacles to interplanetary communication 51, 55, 62–63, 74, 85, 109
Ogden, C. K. and I. A. Richards 66, 68, 77
Ollongren, Alexander 104, 106–108

Peebles, Curtis 13
Peters, John Durham 45–72

Index

planetary detection 30 *see also* exoplanets
Project Blue Book 1, 3, 6–17, 32, 71
Project Ozma 28, 62, 99

radar 8, 15, 18–19, 56–57, 60
radio 51, 53–57, 59–61, 74
RAND Corporation 20
Roberts, Adam 80, 95–96
Robertson panel 6–9
Russian contributions to SETI/CETI 29

Sagan, Carl 29, 34, 103, 120, 128–134
Samuels, David 84–88
Sapir-Whorf 87, 138–139 *see also* language, theories of
science fiction 42, 68, 72–76, 88–97, 119–120, 142–143
SETI (Search for Extra-Terrestrial Intelligence) 1, 4, 11, 27–34, 42, 61–65, 98, 103, 110
solipsism (cf. telepathy) 52, 58, 64, 66–67, 109, 123, 125, 127, 140

Space Policy, origins of 17–27
spaceships 16, 54–61, 135
Spiritualism 49–56, 58, 62, 142
Sputnik 7–12, 22–23, 26
Stephenson, Neal 84–86, 89
Swedenborg 55, 99
Swords, Michael 6–16

technocracy 18–19, 22
technology (overlap of civilian and military interests) 17–25
telepathy (cf. solipsism) 52, 57, 60, 63, 67, 81–82, 84, 97, 109, 131, 140 *see also* mind-to-mind *under* communication
Theosophy, theosophical contributions 35, 39, 41, 49, 54, 56, 127, 129, 133, 142
therapy 65, 67, 69–70, 79
time travelling 63, 75, 97, 132, 139, 143

UFO sightings 3–17, 45

Wells, H. G. 27

Mini Series: Images of Elsewhere
TIMOTHY JENKINS

Vol. I
Flying Saucers: An Introduction

Vol. II
Religion and Science Fiction

Vol. III
Martian Linguistics

Vol. IV
UFO Reports

Vol. V
Alien Sightings

Vol. VI
Images of Elsewhere

www.ingramcontent.com/pod-product-compliance
Ingram Content Group UK Ltd.
Pitfield, Milton Keynes, MK11 3LW, UK
UKHW021838140426
5217IPUK00022B/1508